BINARY PHYSICS

JOSEF ZILBERBERG

Binary Physics / Josef Zilberberg

All rights reserved; No part of this book may be reproduced, reprinted, scanned, stored in a retrieval system, transmitted or distributed in any form or by any means, electronic, mechanical, photocopying, recording, or by any other current or future means without the prior written permission of the author or his representative.

Copyright © 2017 Yossi Zilberberg

Translation from the Hebrew: Academic Language Experts
Contact: yossi@after-school.org.il

ISBN 978-1545203200

Dedicated to my father.

Table of Contents

Introduction 9
The Axioms of General Binary Physics 17
The Rules of Private Binary Physics 19
The language of Binary Physics 21

Chapter 1
First Axiom:
"Existence" and "Placeholder" 27

Chapter 2
The Second Axiom:
The Binary Field (The Universe) 31

Chapter 3
The Limits of Imagination 43

Chapter 4
The Third Axiom:
The Existence Algorithm 59

Chapter 5
General Binary Physics
Versus Private Binary Physics 67

Chapter 6
The Private Existence Algorithm
(The Ten Rules of Physics) 75

Chapter 7
The Fourth Axiom:
The Initial State 77

Chapter 8
The Fifth and Final Axiom:
The Queue 81

Chapter 9
The Dimension of Depth..89

Chapter 10
The Fourth Dimension:
The Dimension of the Queue..95

Chapter 11
The Fifth Dimension:
The Dimension of the Depth Plane 101

Chapter 12
The Sixth Dimension:
The Dimension of Change in Depth 121

Chapter 13
The Seventh Dimension:
The Dimension of the definitions of the initial state 127

Chapter 14
The Eighth Dimension:
The Dimension of the definitions of the Existence Algorithm ..129

Chapter 15
Resolution..131

Chapter 16
The Obstruction
The Fundamental Force..143

Chapter 17
The Nature of Motion 159

Chapter 18
Systems..183

Chapter 19
Change (Energy)...215

Chapter 20
The Birth of Substances and the Nature of Mass......................248

Chapter 21
The Illusion of Speed and Time
and the Source of Relativity 298

Chapter 22
The Secret of Gravity..321

Chapter 23
Time Travel..335

Chapter 24
A Brief History
of The Consciousness..345

Chapter 25
The Will 369

Chapter 26
Being, Feeling and Experience...381

Chapter 27
Is God is a Binary Force?...387

Chapter 28
Calculating the Resolution 393

Epilogue 397

Main Sources of Influence and Bibliography...........................399

Introduction

One particle. One force. One theory.

Binary Physics — The Theory of Everything.

I was once told that if it is impossible to summarize the key insight of a scientific book in one sentence, then there is no point in reading it. So in order to spare you pondering whether what they told me is correct or not, I will distill the essence of my book in one brief sentence:

> *Time Resolution - "The human brain 'shrinks' on the time axis, different cyclical patterns of a single elementary particle, and imagines it as different particles in space."*

Understand this and you will understand all about Binary Physics.

For every particle, photon, electron, anti-electron, proton, anti-proton, neutron and all their associates, and of course for all materials composed by them, there is a fixed "DNA" pattern based on the frequency of the appearance of a single elementary particle along the time axis.

In Binary Physics, the time axis is a physical dimension, entirely akin to the three spatial dimensions, and is called "depth". Thus, it dovetails with the three well-known spatial dimensions, i.e. length, width, and height.

Our consciousness interprets the frequency of the appearance of this single elementary particle in depth as different substances in space.

The manner in which our consciousness interprets the depth at low resolution, as something flat, as the present, can be illustrated by interpretation of numerous pixels on a television screen at low resolution as if they were one essence. As we look closer at the screen, we realize that all the shapes and figures are actually constructed from the same particle, the pixel.

Our consciousness has not yet sufficiently evolved to perceive this, but a process identical to compressing many pixels in space and interpreting them as one essence, occurs along the time axis as well.

Imagine that our consciousness takes every point in space at "100 time units", in which there is a specific pattern of the elementary particle, and interprets them for us "compressed", according to the frequency of the appearance of the elementary particle, as a variety of particles with different properties like mass, magnetism, energy, electric charge, etc.

But just as with the television, if we examine all matter at a high enough resolution along the time axis (to which our consciousness is blind), we will see its "DNA" pattern - the pattern of the elementary particle of which it is composed.

Our consciousness will interpret a point in space with the depth pattern 110000111 as a substance (a particle) different from that of a point in space with the depth pattern 001100000.

It is clear that "100 time units" have been chosen here for illustration purposes only. According to my calculations, which I present later on in the book, the relative time unit of consciousness—the second (when it is measured on the earth's surface on a clock which is at rest) — consists of 1.343×10^{50} turns (which is an objective, fundamental unit of the dimension of depth). In other words, in one second of our consciousness, the human brain compresses 1.343×10^{50} units of depth (in which the elementary particle appears or does not appear) and presents them at low resolution as one unit—one essence—one substance/particle with specific properties.

In short, there is no such thing as a "present" time. Our consciousness compresses a tremendous amount of "time units" ("turns" – the units of depth) over which the elementary particle is distributed, and presents them to us as a single shallow reality, in which there are "different" particles, where each particle is an interpretation of a different frequency of the elementary particle in the dimension of depth (the time axis).

I realize that this is difficult to grasp. Our consciousness is "blind" to depth. It conveys to us that it sees many types of particles, viz. Photons, electrons, anti-electrons, protons, *etc.*, at the "present" time, despite the fact that there is no such time. And there are no such particles.

What depth pattern does mass have, and what pattern does energy have? What pattern does a particle have, and what pattern does an anti-particle have? What pattern does an electron have, and what pattern does a proton have? I will answer all these questions in detail in Chapter 20, *The Birth of Substances and the Nature of Mass*. This is the most important chapter in the book. So if you want to read only one chapter - read this one.

If you can succeed for a moment in using your imagination to look beyond space into the dimension of depth, you will understand the rules of Binary Physics in their entirety.

From the instant that I grasped this new insight, I understood that it can be used to explain how all the rules of physics can be defined by means of a single simple computer algorithm. The insight that it is possible to explain all of the properties of materials based on the pattern of the appearance of a single particle—this is the key that allows the physical world to be translated into a binary one.

In the digital world in which we live, following the Digital Revolution, it did not seem logical to me to continue explaining our reality in terms of Newton's concepts that were defined hundreds of years ago, a long time before the Industrial Revolution.

I decided to have the courage to proceed in a way that people who are wiser and better than me have already stridden and to try and formulate my own "Theory of Everything", a theory that will speak the language of our generation, the digital language.

I set a goal for myself to construct a simple theory based on a single elementary particle and a single force, a theory that can be explained by an algorithm, which can be formulated in a few lines.

As you can see, the goal looked ambitious and unattainable…But, nonetheless, I felt a real need to go ahead and commit the physics ideas that constantly flooded my mind to writing. It occurred to me not to give up, and to write at least the fundamental principles.

Thus, my theory was born. Binary Physics.

Initially, I defined five axioms that determine the binary boundary of human imagination. Through these axioms we can configure a huge number of universes, and to form a tremendous number of "new physics". Of which understandably the physics of our Universe is one case. These became the Five Axioms of General Binary Physics.

The goal of Binary Physics is to give a unique definition for every observation that we perceive or are able to imagine within the framework of its axioms.

Within the limits of the axioms I defined an algorithm of 10 rules, that constitute the private rules of our universe. This algorithm is called the "Ten Rules of Special Binary Physics".

Binary Physics is formulated in such a way that makes it possible to run it like an algorithm by a computer, and by doing so to define any physical observation.

By means of these five (simple) axioms of General Binary Physics and the ten (no less simple) rules of Special Binary Physics, I started to interpret all the classical physics concepts, one after the other: protons, electrons, anti-particles, electromagnetic radiation, spin, molecules, electrical attraction, gravity, the speed of light, space-time, mass, energy, force, frequency, Einstein's Theory of Relativity, Newton's laws, and more…

When I commenced my research, a wonderful world started to appear before me.

A world in which there is only one particle and only one force, speed and mass are illusions, light and matter are identical, the future influences the present no less than the past, and the past is constantly changing, the Big Bang is still going on, God is a system and Newton's First Law is reversed…

Despite the eccentricity and unintuitive nature of this world, it is possible to say that one thing about it is certain, it is incredibly logical. If you logically follow three most fundamental, basic concepts that allow for human consciousness—Existence, Non-existence, and Change—and try to describe a wide range of physical observations by them, you will have no choice other than to reach conclusions identical to those of Binary Physics.

Many possibilities for future research and technological development are concealed in the world of Binary Physics. The fact that for all matter there exists a specific pattern in depth ("DNA") that is based on a single particle, could allow for incredibly easy duplication of substances and complex structures.

The understanding that the physical depth pattern creates what we call a "magnet" in space would permit development of many kinds of "magnets" in depth, which will make it possible to hover over any matter, *etc.*

I am not a mathematician. I am a physicist. I tell the story of the nature. Like any story, it can be conveyed in many languages. Despite my broad knowledge of math and my deep love for it, I decided to write the book in a human language. I wanted Binary Physics to be clear to everyone, even to the vast majority who are not experts in physics or mathematics. Therefore, I decided to reduce the use of the mathematical language to a bare minimum.

Another decision that I made was to write Binary Physics in a descriptive, rather than computational, manner. In other words, to describe the "Existence" (the elementary particle of Binary Physics) as a particle that definitely exists and not as a mathematical value of a cell that performs calculations. The presentation of Binary Physics in a descriptive manner influences only the way in which the reader will visualize the insights of Binary Physics. The same insights could be presented in a computational manner with no essential difference. As a physicist, I decided to describe the elementary particle "Existence" as having a real existence, because I felt that the physicist's role is to tell the story of the nature in a way that it will be as easy as possible for human systems[1] to visualize the story in their minds.

It is understood that there is still a lot of work to do... There are still innumerable concepts that have to be defined, and I assume

---

1. I refer to human beings as "human systems", in order to remind us that there is no physical difference between us and any other physical system.

that even those that I have defined can be defined even better...

I believe that, on the basis of this book, people who are more qualified and talented than I am will arise and write the complete tale. It's hard to know what your grandchildren will study. But if they study physics, they will definitely take this course at university. Binary Physics.

I am excited about publicizing these ideas, because this is the greatest thing that I have done or will do in my small world.

Yours truly,

    J. Zilberberg

The Axioms of General Binary Physics

First Axiom – The Existence and Placeholder
"There are two essences, Existence and Placeholder. One can change into the other."

Second Axiom – The Binary Field (The Universe)
"There exists a Binary Field of (at least) 8 identical dimensions, which is divided into cells, and each cell contains one of the elementary essences."

Third Axiom – The Existence Algorithm
"There exists a continuous algorithm, which can be understood by an internal observer, which determines when Existence will convert to Placeholder and vice versa."

Fourth Axiom – The Initial State
"There exists an initial state, in which and prior to which, the values of the cells are not subject to the Existence Algorithm."

Fifth Axiom – The Queue (The depth dimension)
"For every cell, there exists a queue, and in every queue, the value of the cell is determined anew, according to the Existence Algorithm."

The Rules of Private Binary Physics (The Private Existence Algorithm)

The rules which can be used to uniquely define any physical observation in our private universe. The universe was created in an algorithmic language, and in this language, we should formulate all the rules of physics. The mathematical language derives from the algorithmic language.

First Rule – The Rule of Autonomy
"The passage of an Existence to another cell in the following queue is determined in an autonomous way, only according to the value of the cells that surround it in the current queue and Rules 2–10."

Second Rule – The Rule of Proximity
"An Existence can only move to one of the cells that situates in a direct proximity."

Third Rule – The Rule of Non-merging
"An Existence can move over only to a cell with a Placeholder."

Fourth Rule – Conservation
"When an Existence moves over to another cell, the abandoned cell converts to a Placeholder."

Fifth Rule – Uniformity of the Queue
"The Queue is uniform for all the cells."

Sixth Rule – Time (The Symmetry of Space–Depth)
"Every movement of an Existence in space requires it, Simultaneously, to move forward a cell in the depth dimension"

Seventh Rule – Trivial Movement in Depth
"The default state of a static Existence in space, is to move in the next turn toward his direct cell in the depth dimension."

Eighth Rule – Inertia
"The default state of a moving Existence in space, is to move in the next turn to the direction to which it moves or was supposed to move in space, in the current turn, while giving precedence in the depth dimension to the Existence before it in the direction of motion."

Ninth Rule – Collision
"When an Existence is found in a cell next to another Existence, if they have different default motions, the direction of movement will pass in the next turn, from one to the other, in the direction of movement."

Tenth Rule – Gravity (Bypassing in Depth)
"When an Existence encounters an obstruction in the depth dimension, it will circumvent it in a circular movement in a fixed order, to one of the next cells in the depth dimension, while giving precedence to the Existence at the current level in the depth dimension, that his previous in the order. This rule takes precedence over the Eighth Rule."

The language of Binary Physics

Binary Physics - "A theoretical branch of digital physic. The theory aims to explain unambiguously any possible physical observation using one algorithm and one particle"

Depth – "The dimensions of the queue"

These dimensions are completely identical to the three-dimensions of space. Consciousness is blind to the depth dimension and interprets its existence as time.

Resolution – "Imaginary reference that gives an interpretation to a number of cells as if they were one cell, which value is a superposition of the values of all its component cells."

Physical concept – "A concept that describes an existing thing at a fundamental resolution."

In practice, only concepts that are directly defined by the axioms or by one of the Ten Rules of private Binary Physics are pure physical concepts.

Consciousness Concept – "A concept that describes the interpretation of the consciousness in low resolution to a physical concept."

In practice, all known concepts of classical physics, like speed, time, electron, mass and energy, are concepts of consciousness.

Existence Algorithm - "An algorithm that determines the basic movement of the Existences at the fundamental resolution in a specific universe."

System—"A mental concept. A pattern of two or more Existence"

Except for the physical concepts, Binary Physics relates to everything as a system—the system of the electron, quark system, proton system, atom system, molecule system, heart system, human system, automobile system, television system, computer system, society system, country system, Planet Earth system, the solar system, galaxy system and the largest possible system—the system of the universe in general.

System Algorithm (in short - algorithm) —"A mental concept. The regularity of motion of the system at low resolution. Based on a superposition of the regularity of all of the movement of Existence that comprises the system at the Fundamental Resolution."

A Stable System - "A system possessing a pattern that does not undergo self-destruction. A system where the internal order of Existence within it matches the character of the Existence Algorithm such that its pattern is preserved throughout two or more turns (Existence does not "flee" from it, and the ratio of Existence to Placeholder is preserved)."

A Complex System - "A Stable System Possessing internal Obstructions in depth (Mass)."

Durable System - "A stable system that her pattern is built in such a way, that the internal spinning of Existence within it, prevents external Existence from entering into it or causing its decay."

The First Rule of Systems: The Rule of Cooperation - "When one system performs a non-random interaction with a second system, the survival of one of the systems is dependent on the second, or they depend on each other."

The second rule of systems – The rule of meaning - "Each system receives its meaning from the system that contains it, and gives meaning for the systems it contains"

Movement – "A mental concept that reflects a change in the value of a cell from Existence to placeholder and simultaneous change in the value of next cell from a placeholder to Existence."

Obstruction - "A situation in which an Existence can not move to another cell because that cell already contains an Existence."

Force = "Preventing change – The interpretation of the consciousness to an obstruction that is external to the system, either in space or in depth."

Mass - " A mental concept that reflects an internal obstruction in the depth dimension of the system."

Energy – " The number of exchanges of the Existence with Placeholder, in a given space–depth. In short – change."

Electric Charge - "A system possessing a depth pattern that allows systems with complementary depth patterns to spin around it."

The electromagnetic force - "The change that is created in the movement of the Existence as a result of spin in depth of a system possessing one pattern, around a system possessing a complementary pattern."

Gravity - "The influence of the obstruction in depth on the direction of motion of the Existence in space."

Spin - "Repetition in Depth."

"DNA" of matter - "The cyclical pattern of the Existence in depth that constitutes the properties that matter."

Time - "A mental concept that reflects the ratio between the rate of change in depth of one system and another."

The speed of time (the rate of change in depth) = vt – "The number of changes in depth relative to the number of turns."

Speed - "A mental concept that reflects the ratio of change in space between one system and another."

Acceleration - "A mental concept that reflects, at low resolution, the increase in the rate of change in depth, and as a derivative of this, the possibility of increasing the speed in space."

The speed of light – "A mental concept that reflects the movement of Existence from cell to cell in the speed of the response time of the cell. In short - the speed of the response time of the cell"

Vector – "A physical concept that points the cell which an Existence is supposed to advance in the next turn, in the absence of an obstruction."

System Input - "A mental concept denoting the impact of the environment on the system"

System Output - "A mental concept denoting the impact of the system on the environment"

Life - "A sub system algorithm that actively responds to input, which harms the system, by means of output that prevents its decomposition."

Awareness - "Algorithm based on the influence of one Existence on other Existences."

Consciousness: "A system that produces an output for itself, and responds to that output."

Parasitic sub system algorithm - "A mental concept that indicates a sub system algorithm that does not assist and/or damage the preservation of the system."

Happiness is an example of a parasitic sub system algorithm.

Free Will - "The mental concept that denotes an autonomic ability of a system to influence her output."

Infinity - "A mental concept that indicating the limits of the axioms"

God - "The system"

Chapter 1
First Axiom:
"Existence" and "Placeholder"

Three things are known to human systems about the Universe:

    1. "Existence"

    0. "Non-Existence"

    10. Change

All the rest is commentary.

The simplest thing that I can say about the universe is: "Existence".

Of course, the "Existence" is meaningless without its undisputed companion, "Non-"Existence".

Two fundamental essences. "Existence" and "Non-Existence". Two essences without which it is impossible to describe anything.

Many times have I tried to think about a description of the Universe by an isolated concept, only "Existence" or only "Non-Existence". Each time I imagined a totally homogenous Universe, it would not be possible to say anything about it besides its very "Existence".

These two essences, "Existence" and its twin brother "Non-Existence", were from my viewpoint the bare minimum in order to describe a universe in which something, a "change", occurs.

Within my theory, I decided to designate "Existence" as a single active particle and its inseparable mate "Non-Existence" as a passive Essence, which would be in every place with no "Existence". The term "Non-Existence" seemed inappropriate to me, therefore I changed it to the "Placeholder".

Thus were the first two fundamental principles of Binary Physics born, the active particle, "Existence" and the passive Essences "Placeholder".

The "Existence" and "Placeholder" are perfectly symmetrical, and it does not matter which one is truly "Existence" or "Placeholder", as long as we maintain the uniformity of the definition that we have chosen at the outset.

The "Existence" and "Placeholder" have a single role. To create change. Therefore, the only property of the "Existence" is the ability to change into a "Placeholder", and the only property of the "Placeholder" is the ability to change into an "Existence".

I also decided to establish a mathematical representation for each of the elementary particles. From now on, the "Existence" can also be referred to as "1" and the Placeholder "0". After I did that, the name of my theory became clear to me, Binary Physics.

Already in the 5th century BC, the Greek philosopher Parmenides held that "What exists, exists. What doesn't exist, doesn't exist.", which is an axiomatic truth. I completely agree with this statement as a basis for understanding the Universe. Parmenides later postulated that everything is fixed and there is no change, because the "Existence" cannot turn into Non-"Existence" and *vice versa*. Of course, I do not agree with the second half of his statement.

Change is an integral part of our being, as well as the "Existence" and Non-"Existence" obviously are.

Since on the one hand I was not able to reduce the number of basic concepts to less than two, and on the other hand I was not able to determine that the "Existence" and Placeholder really constitute the bare minimum (it would be reasonable to assume that there are strange worlds, that I cannot imagine, having only an "Existence" without a Placeholder, or a half "Existence" and one-quarter Placeholder, or one-and-a-half "Existence", *etc., etc.*), I decided to set the following definition as the first axiom of General Binary Physics.

The First Axiom of General Binary Physics:

"There are two elementary essences – the "Existence" and the Placeholder. One can change into the other."

I think that the basic ambition of physics as such, as a science which purpose is to describe the nature, is to try to achieve this goal with the minimum number of essences that the human mind can allow.

Having laid down this axiom, I decided that henceforth I would not be tempted to add more particles, and I would refrain at all costs from going back to the jungle of particles of classical physics. My theory will seek to explain all the observations in the nature with the help of only the "Existence" and "Placeholder". Adding additional particles is unnecessary, for according to the axiom, every additional one will be defined in any case by means of these two elementary particles.

Practically speaking, because the Placeholder is a passive essence, which role is to fill every place where there is no active particle of the "Existence", and in the sense that we treat a particle as something that exists, there is only one particle in Binary Physics, the "Existence". This is how I will refer to this matter from now on.

The Principle of Realism

Since Binary Physics is a combination of the science of physics and computer science, it is committed to the principle of realism. It states that in the basic resolution, every system has defined properties, whose "Existence" or values, are independent of measurements. There cannot be a situation in which a particular system may simultaneously have and not have a particular property. Measurements show us properties of systems, and may affect them. But if a change happens because of the measurement, it does not affect the system's defined properties that existed before it.

Chapter 2
The Second Axiom:
The Binary Field (The Universe)

As my thinking progressed, and together with my research, I realized that I am significantly constrained as an observer, in that I am actually part of the system that I study, and moreover I am also confined by it.

Certainly my work would have been easier if I had the possibility of viewing the Universe from outside, and without the effect of the rules that apply there.

But since I have been given no choice in the matter, there is no other way than to set down a number of assumptions that I do not examine. Assumptions that constitute the limit of the imagination of an internal viewer in the system that confined by it, and because of that can't be proved by him.

Axioms.

Just as I decided to limit myself to one particle, the "Existence", so I decided to reduce the number of the axioms of Binary Physics to a bare minimum, and to try to explain as many observations directly from the model, and not external to it, by Adding axioms.

I came up with five axioms that were required for Binary Physics to come into existence and start moving.

Perhaps in the future, it will be possible to reduce one of the axioms, or maybe even all of them. I have not succeeded.

We encountered the first axiom previously, and we required it in order to define the elementary particles "Existence" and "Placeholder".

Now we are in need of the second axiom: The axiom of the Binary Field.

The Binary Field is the place (to the extent that it is possible to call it a "place"), where Binary Physics occurs.

The Binary Field is divided into cells. Each cell contains one of the two elementary particles, "Existence" or "Placeholder". No cell is considered "empty".

Binary Physics assumes that our entire Universe is one large Binary Field. In the universe of Binary Physics, there is no region that is outside of the Binary Field. It would be more correct to say that, if there is a region outside of the Binary Field, then this region will have no effect on the Binary Field itself.

Many times throughout the book I will call the Binary Field by its popular name, "The Universe".

"Time" is an inseparable part of the Binary Field, and it is constructed of cells that contain either the Existence or the Placeholder, exactly as does "space".

As opposed to classical physics, where the future and the past are in parallel universes or at least are not a standard part of the universe found in the present, in Binary Physics all "time" from the beginning to end is an integral part of a wave of Binary Fields in the universe. There is no difference between Binary Fields of the past, present and future.

We will learn later that in Binary Physics the only difference between time and space is our psychological illusion as observers inside the system and our brain's interpretation of the nature of the flow of the Existence within Placeholder. Our brain interprets one flow pattern type as "time" and a second one as "space". In both cases, we speak about the very same process.

Since in Classical Physics the concept of time is separate from that of space (at least in physics that preceded the Theory of Relativity) and since in Binary Physics we deal with an aspect of the Binary Field that is identical to space, I decided to call the Binary Field, which is identical to time, by the name "depth". In doing this, I disconnect the "depth" from all of the associative thinking that links it, for us human systems, to the concept "time".

Our Universe has three binary dimensions that represent space, viz. height, width, and length. There are other binary dimensions (to be precise, three of them) that represent depth, and finally two more dimensions that belong to General Binary Physics, which allow for the existence of other universes with rules of physics different from our own, or an initial state that is different from ours.

The concept of depth is one of two fundamental concepts of Binary Physics, the second being the concept of resolution. Therefore, I will later devote a separate chapter to it, in which I will explain it and the necessity for the various dimensions.

The eight dimensions of the Binary Field are the bare minimum, which we need to describe every observation that we see in the present, past or future in our Universe.

Other universes may require more or fewer dimensions, depending on the number of connections that each cell possesses. We will see later that the number of connections of each cell is what determines the number of dimensions in that universe.

Ultimately, the Binary Field must be able to describe every imaginable observation.

> *The rule of thumb is: If I can imagine it, Binary Physics must be able to give it an unambiguous definition within the framework of its axioms.*

There is no meaning in defining the maximum or minimum number of cells per unit area in a universe, only that we determine that nothing exists between the cells.

It is possible to put this differently. Since the only thing that counts is the way each cell affects other cell, even if there is some sort of existence in the space between the cells, then this existence has no influence on the universe, and therefore is physically meaningless. Each significant region is defined by means of the cells.

The unit of the Binary Field is the cell, which can be abbreviated "cl".

The cell is the basic unit of Binary Physics. This is the only objective measuring rod in Binary Physics. All other units are defined in relation to it, and we will discuss them as relative units. The cell unit is an absolute measure.

What is the size of the cell? The cell is an axiomatic concept, and therefore its size is also axiomatic and cannot be defined or provided within the system. Practically speaking, the size of the cell is meaningless. Once we have defined that each cell can contain only one of the elementary particles (Existence/placeholder), and this elementary particle determines the meaning of the cell, there is no importance or significance of the cell size. Just as an example, a "large" cell that contains the Existence has the same meaning as a "small" cell that contains the Existence. Therefore, there is no point at all in speaking about the size of the cells. If at present it contains the "Existence", then this is its significance, and if at present it contains "Placeholder", then this is its significance.

All the dimensions of the binary field —length, width, height and depth —are measured by the units of cells. Therefore, comparisons within the universe are always relative in relation to the cell units.

For example, when an internal observer (within a universe) determines that the distance between him and the table is 5 cl (we write that distance s = 5 cl) he thus defines an objective and absolute figure that is to be agreed upon by any observer within the system or external to it whether the observer is static or in motion.

On the contrary, when an observer within a universe says that the distance between him and the table is 1 meter, he refers to a relative measurement. This is because the size of the meter is subject to the interpretation of the observer's mind. When I will speak at length on the illusory concept of speed (speed does not exist in Binary Physics), I will also explain why a 'meter' that moves "faster" is shorter (in the objective terms of cell units).

Because in Binary Physics the concept of time is totally identical to the concept of space— both are binary fields, built the same way— it is possible to measure time in units of cells (or if you will use the concepts of Binary Physics, the depth is also measured in units of cells).

In Binary Physics there is symmetry of space–time.

We will discuss the matter in detail in the chapter devoted to the concept of depth.

The Second Axiom—The Binary Field (The universe)

> *"There exists a Binary Field of (at least) 8 identical dimensions, which is divided into cells, and each cell contains one of the elementary essences."*

When I continued to investigate, I discovered that the axiom was only the beginning of the discussion of the field—the universe.

I had no alternative but to set down the very existence of the field in an axiom. But what form would it take? Spherical? A quadrilateral? A rhombohedron? What would be the angles between cells? What would be its size? Does it have an end?

The axiom does not define all these things, since there is no reason to define axiomatically insights that can be derived from observation by means of research and contemplation.

As long as no other definition is required from the observations, and for the sake of simplicity, I have made a number of assumptions about the private field of our Universe:

The first assumption is that we are dealing with a cubical universe, in which the cells are arranged at 90 degrees to each other. This assumption is based on the three spatial dimensions that we know from observations. The three spatial dimensions allow us to define the observations better.

Is there a meaning in different angles between cells?

It is worth noting that there is no meaning in the angle between cells, precisely just as there is no meaning in their size. Once a cell is connected to another cell, it affects it regardless of the angle (if it is possible to call it an angle) at which they are connected. The type of connection between cells is what is significant: how many cells are connected to each cell, and in what order. The angle has no significance.

Binary Physics does not allow "half-influence" or "43° of influence" on the cell. The Existence Algorithm needs only to determine which cell will influence another cell. Just as there is no significance to the space between one cell and another, whether the spacing is small or large, so there is no significance in the angle.

From the same considerations, there is also no significance in the structure of the universe if it is spherical, square, or rhombohedral. The number of connections of each cell, the order of the connections and the Existence Algorithm (about which we will learn later) are what create our illusion of the space structure.

I call the space in the universe by the name "the cube of space".

If we exist in the cube of space—how do we see circles? The circle is an illusion of our mind as it contemplates the cosmos at a low resolution.

The "circles" of Binary Physics are called binary circles, which are essentially an approximation of a circle, just as the circle that we see on the computer screen is actually an approximation of a circle by a lengthwise and crosswise weave of pixels.

I have not calculated this, but I believe that the number π is not infinite, but rather its length is proportional to the level of resolution at which we, human systems, observe the Universe.

In the chapter on the nature of motion and the illusion of speed, I will explain the source of rotational motion, which is so common in our universe, and what makes us obsessed, to an extent, with circles.

The Meaning of the Spatial and Depth Dimensions in Binary Physics

The number of spatial dimensions and depth are derived from the number of cells that are connected to each cell. Each additional connection that we add to a cell will create, at the low resolution of our consciousness or the like, a sensation that a new dimension has been created.

The number of cells that are connected to each cell directly influences the quantity and nature of change within the universe. As there are more connections between cells, thus we can see more complex patterns of change. The brain of human systems interprets patterns of change that result from the number of connections between cells as dimensions.

If each cell in a cube of space connected only to depth cells before it and after it, then we have a point in space, with 0 spatial dimension.

If every cell in a cube of space is connected to two cells in the cube of space—for example, the cells to its right and to its left in addition to the depth cells—then we have a linear space, with 1 spatial dimension. Of course, when I say to its left and to its right, this is just on the reader's level of representation. There is no significance to the true physical location of the cells.

If each cell in a cube of space was only attached to four cells in his cube of space—the cells to its right, left, front and rear (again, the concepts of "front" and "rear" are only intended to help the reader visualize) — then we have a plain space. 2 spatial dimensions (two-dimensional space).

If every cell in a cube of space is connected to 26 cells in the cube of space (the cells on the right, left, front, rear, below and above, and all the combinations of the diagonals of these cells), then we have a cubic space, i.e. three-dimensional space.

Once we attach other cells to a cell, beyond the 26 space cells, we will receive additional dimensions.

Therefore, when we connect the cells in one cube of space to cells of two cubes of space (one "in front" and one "in back"), we create another dimension. This is the dimension of depth.

The last situation occurs in our private Universe, and therefore we will deal with it in the framework of our private physics. In our private Binary Physics, every cell is connected to 80 cells: 26 in its cube of space, 27 in the cube of space in front of it, and another 27 in the cube of space behind it.

General Binary Physics allows for the existence of many other universes. Therefore, in the framework of General Binary Physics, there may be universes in which each cell in a cube of space is connected to more than 26 cells in its cube of space, and every cell in the cube of space is connected to more than two other cubes of space. Thus, for example, we can have a universe with four dimensions of space and two dimensions of depth.

If for example, each cell is connected to more than two cubes of space in depth, it would form a universe in which "time" is parallel with each connection of each additional cube in space. From observations, we see that this is not the situation in our Universe. When I go with my good friend Pie (which is also the dog of my son Jonathan) for a walk in the park, he continues to walk alongside me even when minutes pass. If there is a situation of parallel time or parallel depth, in the language of Binary Physics, then I would have to see that Pi disappears at times, since he goes to one cube of space in depth (in time), while I proceed to the cube of space of a second depth.

Another example, if each cell had more than 26 connections to the cube in space, we would see geometric shapes possessing four or more dimensions in space. Since I, a human system, is located in the universe with 26 connections, my consciousness is unable to imagine a geometric figure of 4 spatial dimensions. It would certainly be interesting to pass for a time to a universe, where each cell is connected to 107 other cells or 134 cells... From the aspect of General Binary Physics, it is indeed possible.

Size of the Universe

My second assumption about the Universe, without any contradictory observation, is that the Universe is so large, to the extent that, for the needs of all calculations, it can be assumed to be infinite.

At the same time, it should be noted that the concept of infinity does not exist in Binary Physics. Binary Physics is a physical theory that gives clear and unambiguous definitions for all observations, within the limits of the axioms. We cannot imagine the concept of infinity, and its existence (if one can speak of such a concept in terms of existence) is outside the limits of the axioms.

The General Point of Reference

As does any other field, the Universe also has its point of origin, where the axes intersect.

This is a very unique point in the Universe, since all objective–absolute physical reality, which is not relative, is measured in relation to this point.

Therefore, Binary Physics calls this point "the General Point of Reference", because in Binary Physics, depth (time) is a binary field that is identical to the binary fields of all the other dimensions, such that the General Point of Reference is also the General Point of Reference of time, i.e. the point of the origin of time.

The General Point of Reference is the point of origin of all of the eight dimensions of General Binary Physics.

General Point of Reference—x_0, y_0, z_0, $d(4)_0$, $d(5)_0$, $d(6)_0$, $p(7)_0$, $p(8)_0$
x: length
y: height
z: width

d4–6: dimensions of depth
p7–8: dimensions of Definitions of the General Binary Physics

Each dimension possesses a size large enough to allow assuming that it is infinite, without impinging on calculations; but also has a point of origin, which is the General Point of Reference.

The distance from the General Point of Reference is measured in cells, and serves as the objective and absolute measuring rod for describing reality.

When it is determined that an event occurred at a depth of 10cl from the General Point of Reference, in terms of binary physics, it would be an objective and absolute measure of when the event occurred. Unlike when I say that an event occurred about 10 seconds after the Big Bang, and use a relative measure. This is because the latter case is a relativistic idea to which different observers can give a different interpretation. The length of the second on the watch of a pilot flying at high speed will contain fewer cells units than a second on a watch of an observer traveling at low speed. The length of one second in "cell" units varies from one observer to another. On the other hand, two observers that say that 100 units of cells passed between one event and the next will agree to this, regardless of the speed at which they are traveling.

In the same way, when we determine that a particular event occurred at a height of 10cl from the General Point of Reference, this would be an objective and absolute measure of the location of the event. On the other hand, when I say that the incident took place approximately a meter away from the point of the Big Bang, I use a relative measure. Even the meter is a relativistic concept, to which different observers, traveling at different speeds, will give a different interpretation. The length of one meter in cell units changes from one observer to another.

Chapter 3
The Limits of Imagination

How can we do what is impossible to dream of?

Certainly you are asking yourselves, "How is imagination connected to such an exact science as physics, and how does it occupy an important place in Chapter 3, in the middle of a discussion of the axioms?"

The answer to this is simple: The axioms are the limits of human imagination.

The ability of human systems to understand the regularity that dominates the system that contains them—the system of the universe—is derived from the imaginative ability of human systems.

By means of the five axioms, and after the discovery of the concept of "time resolution," we are able to explain the regularity of every physical observation within the system of the universe. At present, the axioms constitute the absolute minimum of assumptions, which are unprovable from within the system, and are required in order to explain its regularity. If we forego the need for one of the axioms, we will be unable to explain all the observations that we see in the system of the universe.

Employing our imagination and our brain's subjective interpretation of our Universe is an integral part of physics in general, and Binary Physics in particular. It is impossible to make any physical statement about the Universe without referring in a significant way to the instrument by means of which we observe the Universe, the consciousness of human systems.

In fact, we cannot investigate the physical reality directly, but only through the imagined reality that our consciousness shows us. Therefore, a primary aspect of physical research is to understand the imagination, its limits and limitations.

We will see later that the human imagination is the bridge to understanding the difference between the real, binary universe and the colorful and beautiful universe that our consciousness displays around us.

We will see how complex phenomena such as color, force and speed can be explained using simple algorithmic models.

In the future, will it be possible to break the boundaries of the axioms?

Breaking the axioms would negate the theory of Binary Physics, and would require writing new physical theory instead.

For example, if it would be possible to explain the regularity of the universe, with less than two entities – Placeholder and Existence— but rather it will be possible to manage with 1½ entities, certainly there would be a need to write all the rules of physics from the beginning. In this situation, even the name "Binary Physics", which indicates physics that is based on two entities, 1 and 0, will no longer be relevant.

The limitations of any system are necessarily derived from the system that contains it. For example, a computer program, however sophisticated, will not be able to run faster than the processor of the system that runs it.

Thus, the ability of the imagination of human systems is doubly limited.

It is confined firstly by the limitations of the human system itself, and secondly by the limitations of the universe system that containing it and defining it.

There are three possibilities:

The first possibility: The axioms, as they are known today, constitute the maximum understanding and imagination that the universe allows for an internal system that is defined by it. In a situation such as this, it does not matter how much human systems will investigate, and how much they will develop evolutionary: they will never be able to exceed the confines of the system that defines them, and thus the axioms will not be broken.

The second possibility: The axioms, as they are known today, do not constitute the maximum understanding allowed by the universe, but they constitute the maximum of understanding that human systems allow. In such a situation, only continued evolutionary development of human systems will enable pushing the limits of the axioms.

The third possibility: The axioms, as they are known today, do not constitute the maximum understanding allowed by the universe system, and are also not the maximum of understanding that human system allows. In this situation, by means of continued physical research by human systems, it will be possible to push the limits of the axioms.

In any case, once an internal system develop within the universe system, and will succeed to identify the regularity of the system that contains it (the system of the universe), from the very definition, no other system will be able to develop within the universe that will deviate from this regularity.

The Axioms – The Limits of Imagination and Understanding

Binary physics is based on the assumption, that the first possibility above is the correct one, and that the axioms as they are known today constitute the limits of the system of the universe, and define it.

The reason that I make this assumption is that it is the most limited of the three. Without any evidence to the contrary, there is no reason to make wider-reaching assumptions.

Currently, Binary Physics (after the discovery of the resolution of time) allows, within the framework of the axioms, formulating an unambiguous regularity for every observation. I even extend this statement, not only to physical observations, but even to physical imaginations of human systems. For example, there is no science fiction movie, however absurd, which presents a reality that cannot be translated, within the framework of the axioms, to an algorithmic-binary language.

It is fitting to note that Binary Physics does not assert that there is no existence beyond the axioms, but rather just the opposite. Binary Physics also does not claim that the axioms will not be breached in the future; but it does claim that, according to what is known today, the axioms are the limits of imagination of human systems, that are confined by the system of the universe. Binary Physics asserts that there exists a reasonable possibility that there are limits to the imagination of human systems--just as far as is known today, there is a limit to the maximal speed of our system.

Another reason to assume the first premise is a practical one. We should make every effort to define the regularity of the universe by means of the capabilities of our imagination and understanding as they are today. There is no practical reason to wait for a new discovery or breaching of the axioms... If the axioms will be breached in the future, certainly there will be those who will bother to reformulate the theory...

> *The axioms are the five basic Lego blocks, known to us today, which we can build and configure with them, everything seemed in observations.*

Hence we can determine that as long as the restrictions of the axioms have not been broken, they define the boundaries of knowledge and imagination of human systems in particular, and of every creature (however intelligent it may be) that is defined by the system of the universe, in general.

As long as the boundaries of the axioms have not been broken, we can assume that all knowledge beyond them is reserved only for an external observer outside the universe, who is not defined by it.

As internal observers confined by the system, we can't understand what occurs between cells, what is half Placeholder, what a non-binary universe looks like and what is infinity...

What is the Binary Field (the Universe)?

A dream? A giant computer? The mind of an intelligent creature? These questions are beyond the scope of the Second Axiom, and, therefore, as internal beings that are confined by the system, we cannot know. Binary Physics, as an exact science, and we, as human systems and internal observers in the system that are confined by it, are not able to understand how the Universe works from outside. The only important thing, from the aspect of physics, is that within the system and in the framework of its axioms, we will decipher the physical rules, and that there will not be any observation that Binary Physics will not be able to define unequivocally. Gaining a complete understanding of the regularity of the Universe rules is a complex task in itself, and it seems that we, as human systems, will have to settle for, at this moment, with this ambitious aim.

I think that awareness of what you cannot know is one of the most important things you can know.

Walt Disney said that if you can imagine something, you can do it. He was right, apparently. He just forgot to mention that there are lots of things that we simply cannot imagine.

Altshuler, the father of Systematic Inventive Thinking, claimed that before the solution to any problem, we must define the "closed system" in which it operates. Only after we understand the limits, can we find the solution within the boundaries of the "closed system". For example, if you are on a desert island and you have to make fire—the desert island boundaries are your "closed system". There is no point in trying to think about lighting fire with a cigarette lighter, when such a mean is not at all available. You have to light fire only with things that exist on your island.

The axioms are the boundaries of the island of physics.

Is there room for imagination in the closed system of Binary Physics?

Definitely so. Imagination is the human system ability that has evolved to allow us to anticipate situations, and thereby increase the chances of preserving our system.

The number of possibilities that Binary Physics offers is so great in relation to the ability of the human mind, that even if all people in the world together imagine something different every moment in their lives, they will not reach the very edge of the number of possibilities. It seems that the imagination, despite the physical limitations imposed on it, will still serve the human systems for many years as a tool for survival.

An example that I would like to give to illustrate the limits of human imagination that the system of the universe sets us, is the "folder of all the images".

I create a folder on my computer that contains all the possible pictures of our Universe.

I do this by means of a simple program.

The computer program produces images by systematically turning on, turning off and painting pixels on the screen. After every change, it saves the image. When it finishes saving all the images that are based on all possible combinations of turning on, turning off, and coloring pixels, it completes its work.

An average 5 megapixel picture taken by an iPhone is 2000 pixels wide by 2500 pixels in length. That is, a total of 5 million pixels. The number of combinations you can make with 5 million pixels, where each can illuminate in four different colors (three primary colors + off mode, which is black) is tremendous. But it is certainly finite. With 5 megapixels of an iPhone, you can take any picture that the human eye can see, from a page of a physics article, to the Eiffel Tower. From the birth of a baby to a daily newspaper.

If by means of a program, we form all the possible combinations of pixels, we have essentially created every picture that the human eye has seen or will ever see. All human knowledge that was or will be, every formula, every newspaper headline, every picture of a child that was or will be born, it's all there.

Every book, every poem, every painting – everything is already in this file. Any writer / any poet / any painter did not invent anything. All has been foretold. No poet can versify, and no writer could write a book that isn't already there.

There is even a picture of you, the reader, while you are reading this book, as well as a picture of you standing naked in front of a mirror. I promise not to peek…

This folder shows all that human systems saw in the past, and all that they will see in the future. Not only them, but also all sentient beings that will replace them. There is no possibility, even theoretically, of creating an image, at the low resolution at which human systems see the Universe, which is not in this folder. Therefore, this folder serves as the boundary and the end of the developed imagination, which we have been granted.

Only an observer who is external to the system, who is not confined by it, can imagine "pictures" that are beyond the axioms (if it would be possible to call the strange things that he sees as "pictures").

As a side comment, I would like to note that there is no reason to treat the example as a folder of all the movies, because it essentially constitutes a collection of all images, or as a set of all three-dimensional images, which constitutes an assembly of pairs of pictures from the file containing all pictures…

It is not by chance that with creativity and free thinking, we have decided to describe the Universe in our imagination by the principles of Binary Physics.

We have, as human systems, a false sense that our imagination is limitless. Therefore, we believe that we have chosen to describe the Universe using the rules of Binary Physics. Impressed in us is a sense that there are many ways to describe the Universe, and we have chosen to describe it using Binary Physics.

We should not be surprised that we succeed in computing everything, and the hot thing today in technology is the "Internet of Things". It is not that we choose to compute everything out of all kinds of possibilities that have been presented to us. Binary thinking, that has allowed us to compute all that is around us, is an inherent property of the Universe, just as flammability is a property of wood, in order for us to be able to set it on fire…

The reality is that there is no other way to describe the Universe. We and the entire Universe around us is designed and configured according to the rules of Binary Physics, and therefore we have no reason to be surprised that when we investigate, we succeed in describing the Universe precisely by means of these principles.

I will give an example. Suppose we created a universe made of Lego. We would allow the new universe that we created to develop. Since we are talented creators, we constructed our Lego universe in such a way that, over time, intelligent beings would develop in it. Let's call them "Lego People". The Lego People would start to investigate the universe, and slowly they would start to reach a number of conclusions about the rules of their universe. They certainly would call this collection of conclusions "Lego Physics". One of the fundamental principles of their Lego Physics would be that there are several elementary particles in their universe—a block with two pins, a block with 4 pins, and a block with 6 pins. I assume that the Lego People would live with a feeling that they chose, using their intelligence, to describe their universe in such an inventive way by means of the theory of the Lego blocks that they created. In effect, our Lego People would have no other way of describing their universe except in Lego terms. They, exactly like ourselves, are internal observers in the system and are confined by it. They are made of Lego, and therefore they know how to define their universe only in terms of Lego.

Ultimately, the fact that we are able to define the Universe around us in terms of Binary Physics stems out of the way our Universe was defined from the outset, and not of our creative ability to choose from among a number of methods of definition.

It is reasonable to assume that observers outside our system, that are not confined by it, would know how to define the "Existence" between the cells, what occurs outside the Universe, and how the Universe can operate with an Existence Algorithm that is not binary...

This is exactly the same as we, as external observers of the Lego people, know that their elementary particles are made of plastic or that there is air composed of atoms of several elements between their elementary particles… The Lego people will never become aware of this… they are located in a system of Lego and are capable of thinking only in terms of Lego…

A good example to understand the limits of the ability of consciousness that is internal to a system, is if you will press the system's "pause" button and freeze the entire universe, stop the progression of its queues (just as we freeze a picture on a screen); a consciousness that is internal to a system will not be able to know this.

Even mathematics, the holy of holies of the exact sciences, is not an objective and absolute tool that can describe any situation as we imagine it. Mathematics is a local and subjective instrument that suits describing our Universe only for its algorithmic nature (and at low resolution—it's mathematical nature). Therefore, we should not be surprised when we speak about the reality around us by means of it, any more than we should be surprised that a child born in Israel describes the reality around him in Hebrew…

In an existence external to the axioms (outside of the boundaries of our universe), most of the chances are that mathematics is useless. Just as Hebrew is useless in a small village near Shanghai…

As binary beings, shaped by Binary Physics and living in a universe that is defined by it, we cannot imagine or understand physical reality that is not mathematical.

The algorithmic-binary language is the language of our universe's creation. Mathematics is the result of the algorithmic-binary universe, and not the opposite. The mathematical concepts represent algorithmic processes at low resolution.

Why, then, did mathematical thinking develop first, and algorithmic thinking only developed in the recent decades, if the latter preceded the former? The answer is a practical one. Only in recent decades have computers been developed that encourage the development of algorithmic thinking as a key to understanding complex processes and defining the regularity of systems. Scholars of the past who did not have access to computers and the world of algorithmic knowledge that was derived from them, began to describe the regularity of the world through low resolution concepts that are intuitive —geometry and arithmetic. Mathematics developed out of these concepts.

The development of the Lego game in the universe of Lego People would be expected just as in our binary Universe, we would develop a computer. We developed a computer not because we chose precisely this invention out of the entire variety of possibilities, but rather because according to the rules of physics in our Universe— that what can be developed here.

Systematic Inventive Thinking

Systematic Inventive Thinking was developed by a wise Jew named Altshuler (whom I mentioned earlier).

His main argument was that every problem should be treated as a closed system and within this closed system, "playing" with the components in different ways would bring us the desired solution.

He devised various tactics to help us connect the various components in creative ways.

For example:

- Disconnecting – To take a component from the system and use it in a different way from how it has been used so far.

- Duplicating – To duplicate a component from this system.

- Reversing – To use a component from this system in the opposite fashion.

- Symmetry breaking – To arrange elements of this system asymmetrically (as opposed to our consciousness' instinct, which prefers symmetry…)

- And so on…

According to Altshuler, every invention in the world is eventually based on reassembling previous inventions in different ways.

I totally agree with Altshuler.

Also in our Universe, eventually the only thing that happens is an arrangement of the "Existence" in different patterns every time, where every "new" pattern is based on the patterns that preceded it.

Of course, there is a huge quantity, and perhaps in terms of the ability of our consciousness, an incomprehensible number of patterns that can be arranged.

But, in the end, just like with Lego, even if there are a lot of options, they have a definite end.

Infinity

Binary Physics defines infinity in the following way:

> *Infinity: "A concept of consciousness that indicates the limits of the axioms."*

Once the mind reaches the limit of axiomatic thinking, since it cannot cross it, it calls it by the name infinity.

As for the universe outside of the axioms—infinity—the mind can say only one thing about it: It exists.

The purpose of Binary Physics is to describe all observations in terms of unambiguous concepts, in the framework of the axioms. Each time when we are unable to give an unambiguous definition within the framework of axioms, we reached the limit, and we call it infinity.

The concept of infinity is beyond the limits of our imagination, and as a practical

Physicist, I do not see any point in using it to describe observations. Moreover, whenever I concentrated my mental powers and tried to cross these boundaries, I felt as if I was becoming insane, as if it was planned that we could not traverse them...

In Binary Physics, the letter *e* indicates the greatest value in the Universe. In other words, the letter *e* is largest there, but finite.

A Note in a Bottle

If a consciousness external to the system (I arbitrarily call it "consciousness", because I have simply no other way to describe something external to the universe system if not by means of our system's concepts) leaves us a bottle with a note inside and the note says what is outside our system, it would need to write the note in the language of our system. So that we could understand it. And therefore, in fact, it would not be able to tell us what is beyond our system. This is just like in order to say something to a computer system, we must translate the message into the binary code. Once the message is translated into the binary code, we restrict it to what we can tell through the binary code—this is just like explaining the Chinese language in Hebrew, without using a single Chinese word… In short, I would not count on a message from the outside of the system that it would be able to break out the boundaries of our consciousness…

In order to remain optimistic, it is very likely that you can change the settings of our system from the external system (to change the axioms) and thus of course to change all the physics, and thereby allow the systems to live in a new physics that will be created, to understand things that we are unable to imagine and that we cannot imagine. The disadvantage of this process is that changing the axioms will cause our universe not to be the same universe anymore. Changing the axioms will break the regularity and obviating the physical research… (just as a universe that contains "miracles" finds physical research to be distasteful, since the definition of miracle is something that departs from the scope of abiding to physical rules…).

In a closed system, as permissible and developed as it may be, any progress will always be limited to the confines of that system. For example, if our system is a racing car that, in terms of its technical specifications, optimally allows traveling from Tel Aviv to Haifa in one hour, then any driver, regardless of how much his skills improve from one trip to another and how often he practices, will be limited by the ability of the system in which he operates, the restrictions on the car. A Lego engineer can create from Lego blocks an infinite number of structures at an increasing level of sophistication and complexity. But even a Lego engineer—no matter how much he will improve his structures from one competition to the next—will still be limited by his elementary particles, Lego blocks.

> *If eventually will develop in the system a consciousness that goes beyond the boundaries of axioms, it will not happen because the system allowed this, but rather because the definition of the boundaries of the system— the axioms—have not been correct from the outset.*

But once we have determined that this consciousness transcends the limits of the axioms, we can define it as external to the system, and therefore its imagination is boundless. At least there is no boundary that is derived from the axioms that internally restrict our system as human systems. I set boundaries—there is a limit to its imagination—the limit of its imagination is derived from the system boundaries that define it, even if this system will ultimately prove to be broader than our system as we have defined it through the axioms.

In conclusion: a type of paradox. The concepts are divided into physical concepts (five axioms and ten rules of Private Binary Physics) and concepts of consciousness, these are all physical concepts of which we are accustomed to speak, that are in fact an interpretation of the mind that sees the physical concepts at low resolution.

The paradox stems from the fact that all physical concepts are derived from the axioms, but the axioms themselves are a concept of consciousness. They constitute the boundary of the imagination of us as human systems.

It is entirely possible that a consciousness external to the system not only sees the external system outside the axioms, but also understands the axioms of our internal system in a completely different way. For example, in the internal system of the Lego People, the first axiom is defined as follows: "There is an elementary particle called the Lego block". Only a consciousness that is external to the Lego system and is not confined by it, such as ours as human systems, is able to understand that the elementary particle is composed of plastic, which is made up of atoms, and that the entire goal of this internal system is to provide pleasure for one little boy.

An external consciousness is not confined by our system, and hence it is not subject to its axiomatic restrictions. It therefore has the possibility of imagining and understanding things completely different from those, to which our minds are limited to as sub systems within the universe system, and confined by it.

Hence, the paradox can be resolved only by a viewer, external to the system, who can define the axioms from an external viewpoint.

Chapter 4
The Third Axiom:
The Existence Algorithm

The Existence Algorithm defines conditions for change of the base units in the Binary Field (the Universe). That is, when a Placeholder will change into the Existence and *vice versa*.

I have intentionally written "algorithm", because it gives an impression that it seems impossible to formulate a principle that runs the Universe in a discrete mathematical formula, but rather through a series of instructions, similar to the way we formulate a computer program.

For example, "If the cell contains the Existence, check if the cell next in depth contains a Placeholder. If it does, move the Existence to this cell."

Research in Binary Physics clearly indicates the algorithmic character of the universe. In this context, it is proper to say that the algorithmic character of the universe precedes and is more fundamental than mathematics, and only due to this algorithmic character can mathematical descriptions be generated in our universe.

In any event, the assumption of simplicity will direct us in formulating the Existence Algorithm using a minimum number of instructions.

We will try to reach a point where a simple algorithm generates and provides an explanation for all the complexity we observe. The goal is to formulate the Existence Algorithm in the narrowest possible way that will address all observations.

It is clear that we do not want a long and tedious algorithm, which includes a very large number of exceptions that are necessary because the basic rules do not succeed in dealing with specific cases.

Alongside the assumption of simplicity, we require an additional one: **The assumption of no contradictions:** we assume that the Existence Algorithm works without contradictions and in such a way so as not to "get stuck" or require external intervention.

The Existence Algorithm must not create internal conflicts that will produce a "bug" in the system. The system should flow smoothly without mishaps.

Discovering the Existence Algorithm is the primary purpose of Binary Physics. Binary Physics aims to explore and find a single, simple algorithm that defines how the Placeholder changes into the Existence and *vice versa*, an algorithm from which we can derive any change that was and will be in our Universe.

In order that the research will have a purpose, at the basis of Binary Physics lays the assumption, that there is a system of physical rules to the nature of the change from the Existence to Placeholder, and that the change is not chaotic.

This assumption becomes the **Third Axiom** of General Binary Physics:

> "*There is a continuous algorithm, which can be understood by an observer internal to the system, which determines when the Existence becomes Placeholder and vice versa.*"

An opposite assumption obviates all physical research. After all, if the change is chaotic (in the sense that it is non-deterministic, as is sometimes implied in the context of quantum physics) then there is no reason or way to understand the regularity of the Universe.

Since we as researchers are an integral part of the Universe and are confined by it, there is another implicit assumption, that the algorithm can be understood not only by an observer external to the Universe, but also by an internal observer. This is not a trivial assumption, and when it is assumed by an internal viewer who is part of the Universe, it is even pretentious. Consider the game *The Sims*. A programmer, who has to decompile/disassemble software rules underlying the game, is expected to work hard. If so, how much more difficult will this work be for one of *The Sims* characters that is operated and configured by the very same software?

Another assumption is that the adherence to physical rules is continuous. That is, the algorithm that conducted the Universe in the past is the same one that runs it in the present, and it will continue to run it in the future. Here also, we are not speaking of an obvious assumption. It would actually be nice to see a change in the rules from time to time...

Changing the rules and replacing the algorithm would invoke scores of difficulties in physical research conducted by an observer internal to the system. Deciphering the rules of the Universe that are based on a periodically changing algorithm is an impossible task, because there is little value to an observation that indicates a particular adherence to physical rules, when a moment later this adherence no longer exists...

Currently, as long as the observations do not require us otherwise, we will hold on to the axiom according to which the adherence to physical rules is continual, and the algorithm is continuous. If future Binary Physics students will encounter irreconcilable contradictions, they would consider in any case worth investigating whether our Universe has been developed on the basis of several different algorithms that have replaced each other.

It is possible to determine that even if there are changes to the algorithm, they do not affect our Universe, by treating every such change as if it creates a new universe, which "initial state" (definition of "initial state" is given further on) refers to the moment when a change in the algorithm occurs.

As an illustration, we observe a good deal of evidence when simple algorithms produce amazingly great complexity. My favorite example to demonstrate the power of simple algorithms in creating a complex reality is the *Game of Life*, which was invented by John Horton in 1970.

This is a particular example of a cellular automaton. The definition of a cellular automaton according to Wikipedia, is as follows.

A cellular automaton is a model that has been investigated in computability theory, mathematics and theoretical biology. It features a lattice (or in general, a graph) of cells, each of which has a finite number of states. Time in the model is discrete, and the state of each cell at a given time is a function of its state and the state of other cells (the neighbors of the given cell) at time t-1. This function applies to all cells, that is, all the cells change according to the same set of rules. Each operation of the function on all the cells in the lattice creates a new generation.

The conventional use of the cellular automaton is by means of a computer program, which develops a given situation, given on the basis of simple rules, and monitors developments over time. The first automata of this type were built with the thought of imitating life by means of a computer program, in which each square on the screen indicated a "cell", and therefore they received the name "cellular automata". The main impetus for the study of these automata came from the *Game of Life* developed by John Horton Conway. This is a game which made significant waves in the scientific community and beyond.

The field of cellular automata was established by the mathematician John von Neumann, who argued that the mechanisms of automata can be seen as a model of phenomena in the natural world. The physicist Ed Fredkin claimed a number of decades ago that you can visualize the Universe as a giant cellular automaton.

Parenthetically, I would like to note that the first who conceived the idea of the entire universe as a huge computer was the computer scientist Konrad Zuse, considered by many, as the person who built the first universal computer. In 1969, Zuse published the book "Calculating Space" in which he argues that the Universe is the ultimate computer, running a very simple type of program, called a cellular automaton.

In the language of Binary Physics, the cellular automaton is the Existence Algorithm that runs the Universe. The value of each cell is determined to be Existence or Placeholder, according to its value and the value of the neighboring cells of the Previous space cube in depth (dimension d(4). In the chapter dealing with depth, I will explain the nature of physical time).

Horton's *Game of Life* is played on a two-dimensional board of squares. Each square can be in one of two modes, with or without a living creature, i.e. "alive" or "dead". The player determines the initial state of the squares. From the definition of the grid of squares, for each creature there can be up to 8 neighbors (above, below, to each side, and the four diagonals). From this moment on, the game develops on its own, without the intervention of a player. Horton referred to the beings as social creatures that live according to the following rules:

- Any empty cell that has exactly 3 neighbors comes back to life.
- Any creature that has one neighbor, or no neighbors at all, dies

- from under-population.
- Any creature that has more than 3 neighbors dies from over-population.
- Any creature that has 2 or 3 neighbors does not change (until the next turn).

Depending on the initial state, a wide variety of situations can be modeled during the course of the game from a population of creatures that becomes extinct, through a population maintaining in its static state, to an expanding population.

Horton was surprised to discover that there may be initial states that would cause creatures to just "come to life" and behave over time in a very complex manner.

In the chapter on the limits of imagination, we discussed the Lego People who were able to develop only physics of Lego and Lego games. The Lego People would think that out of all available options, they would have chosen precisely to invent Lego games. We, as observers external to their system, smile and understand that they had no other option. We have designed their universe from Lego.

The above example obligates pondering about ourselves as human systems. Did we invent the *Game of Life* by chance out of many possibilities, or perhaps we "invented" precisely this game because we live in a universe that is defined by cellular automata, and these are the only games that we are capable of inventing.

Are there any creatures external to our system that look at us with a smile when we "invent" the *Game of Life*, and are confident that only thanks to our wonderful minds did we come up with this idea?

My belief is different from the popular idea that human systems belong to a superior race, which is unlimited in their abilities.

My belief says we are part of a system and are confined by it, and, therefore, any invention of ours is necessarily part of the system, and we can learn from it about its character. Our sages have already said: "Explore the world in an eggshell..."

I therefore conclude that if the *Game of Life* is not accidental, but deterministic, then there is reasonable room for the assumption that it is possible to learn from it about the system in which we live, and to conclude that the Existence Algorithm that runs it is based on the principle of the cellular automaton.

In this context, it is proper to note that there are not a number of different "Games of Life", where each one displays a separate mode of thinking (I do not mean variations of the *Game of Life*, but rather in the sense of a model of thinking different from an algorithm that is based on the cellular automaton). As I said, in my opinion, we invented a *Game of Life* based on the cellular automaton not by chance, just as not by chance a Lego person would invent a *Game of Life* based on Lego blocks. The cellular automaton is the basic Lego block of our Universe.

Therefore, as long as this assumption remains undefeated, I take that the Existence Algorithm of our Universe is based on the principle of the cellular automaton.

This assumption becomes the First Rule of our own private Existence Algorithm – Private Binary Physics:

The First Rule – The Rule of Autonomy

> "The passage of the Existence to another cell in the next turn is determined in an autonomous manner, solely according to the value of the cells that surround it in the current turn and Rules Two through Ten."

Of course, there is an exception to the First Rule: the Initial State. Therefore, this state is determined axiomatically, and constitutes the Fourth Axiom of General Binary Physics.

As for the concept of the "turn", I will explain it in the framework of the discussion of the Fifth Axiom that treats it.

When examining the universe we see that the private Existence Algorithm not only exists, but also consists of 10 simple rules that comprise our Private Binary Physics. Throughout the book, it is understood that we will become familiar with them and will understand how they are derived from observations, and explain the observations.

If you run the private Existence Algorithm on a computer for enough turns, and you study it at a low enough resolution, you will get any physical observation that exists in the Universe, including observation of the of human systems with consciousness.

Does the ability of the Existence Algorithm to stand the Turing test (that is, to allow for any kind of calculation that a computer can perform) constitute one of the conditions of its existence? The answer is "yes". If we wish to be more precise, it is fitting to phrase it in the opposite way, the fact that the source code of the Universe—the Existence Algorithm—allows for any calculation that a computer can perform, is the sole reason that after a sufficient number of turns, the systems of computers developed.

Chapter 5
General Binary Physics Versus Private Binary Physics

Although the human imagination is not capable of going beyond the limits of the axioms, it is certainly able to describe, within their framework, different universes, even universes that are very different from our own Universe.

Within the binary axioms, it is possible to imagine a huge number of Existence Algorithms. Each new Existence Algorithm creates a new and unique universe.

General Binary Physics presents a range of possible Existence Algorithms within the framework of the axioms. There is a huge number of options (I try not to use the concept of "infinite", since this concept is outside our ability of comprehension, and that even though the number is huge, it is still finite).

In the framework of General Binary Physics research, it is possible to make many assumptions that may be different from those of our Universe, thus creating different universes, of course, as long as we do not contradict the axioms. General Binary Physics examines how our Universe would appear under different Existence Algorithms.

General Binary Physics research is not concerned only with examining Existence Algorithms, different from that of our

Universe, but also studies different initial states (see the Initial State Axiom below).

In fact, each combination of an existence algorithm and an initial state creates a universe, which is valid from the aspect of General Binary Physics, but different (and at times very different) from the Universe in which we live.

It can be said that each combination of an existence algorithm and an initial state constitutes a "3D printer" for a different universe.

In General Binary Physics, anything we can imagine is possible. I imagine that a green alien is landing here now in the middle of the room and makes me a millionaire. In General Binary Physics, this scenario exists.

Moreover, in General Binary Physics, even what we cannot imagine, but which could be imagined by an internal consciousness that is more advanced and wise than us, is also possible.

If we add, along with the complexity of different initial states and different Existence Algorithms, also a different structure to the universe, in terms of number of connections per cell, which influences the number of dimensions, we will truly receive an unprecedented range of possible universes.

Parenthetically, I will note, that it should not be taken as self-evident that every cell has the same number of connections to another cell. It could be a universe, in which, when you move in space, the number of dimensions changes. It would be interesting to go out of my three-dimensional house and to enter my five-dimensional office, and to have dessert in a café with a point dimension. Although there are definitely days when I finish work and feel like no more than a point in space, I do not take this as an observation or a proof that even in our Universe the number of dimensions in space changes… From the observations in our Universe, I assume that every cell in our private physics is connected to an identical number of other cells.

Our Universe is a specific case of a particular Existence Algorithm, combined with a certain initial state, with 80 connections for each cell. Therefore, the study of our own private Universe is called Private Binary Physics research.

Private Binary Physics is a branch of Binary Physics that investigates the specific Existence Algorithm, which operates our Universe with our specific initial state. This is by means of contemplating observations around us.

I will give an example.

Cell A has the "Existence" and Cell B has the "Placeholder". Now we want the "Existence" to pass from Cell A to Cell B. General Binary Physics allows doing this in two different ways.

In the first way, the "Existence" remains in Cell A, and the "Placeholder" in Cell B converts to the "Existence" as well. That is, the movement does not erase the first "Existence", but only replicates it. In this situation, we will have two "Existences" instead of just one.

In the second way, the "Existence" in Cell A changes into the "Placeholder", and the "Placeholder" in Cell B changes into the "Existence". That is, the movement will imitate the first Existence and will maintain the quantity of "Existence" in the system.

Which of two ways characterizes Private Binary Physics (that operates in our Universe)?

From observing our Universe, we know that when we move an apple to the right, it is no longer in the original place where it stood, but rather in the place to which we have moved it. That is, its existence has "disappeared" from the original place where it was, and now it exists only in the new place.

We also are familiar with the Conservation Rules of classical physics that reflect our understanding of our observations that, in our Universe, existence does not appear out of non-existence.

Therefore, you can determine that our Existence Algorithm has a second characteristic. That is, it states that when the Existence moves from one cell to another, the Existence in the original cell turns into the Placeholder, so that the overall number of Existence in the system is conserved.

We have just ascertained another rule of Private Binary Physics, **the Fourth Rule: Conservation:**

> "When the Existence passes to the next cell, the cell from which it came turns into the Placeholder."

General Binary Physics definitely allows for a universe in which the Existence Algorithm is precisely of the first characterization. That is, when we push the apple, a long line of apples is formed along the entire length of the push...

Certainly it would be interesting to pay a visit to such a universe. This universe would easily solve the problem that students have (as well as billions of other people) to deal with lacking economic means. On the other hand, in every movement of a human system in this universe, hundreds or thousands of that human system would appear...

Another important point about the Fourth Rule: The Fourth Rule does not hold only in the dimension of space, but also in the dimension of depth (the time axis). When an apple advances in the dimension of depth, the Existence that comprises it passes from one spatial cube to the next, rather than being replicated in such a way that a trail of identical apples is formed on the depth axis....

On the basis of what do I say this?

First, simplicity. Without any contradictory evidence, there is no reason to assume that our Existence Algorithm determines a different rule for movement in space from that for movement in depth.

Second, the replication of the Existence in the depth dimension will significantly limit the movement of the Existence in space, in light of two rules that we will learn later.

The Third Rule—the Rule of Non-Merging and the sixth rule – Time (The Symmetry of Space–Depth), except on the frontier of time (this concept is explained in Chapter 11).

Third, since the pattern in depth of the particle is what determines its properties, making the assumption that the Existence duplicates itself upon moving in the dimension of depth, will necessarily cause a change in the pattern and a change in the properties of particles over time. In observations, we witness particles retain their properties over time, and thus they must retain their pattern in depth.

It is understood that in General Binary Physics, a universe can exist in which the Fourth Rule is different in space and in depth.

To differentiate the Existence Algorithm of our Universe from all the possible Existence Algorithms, we will call it the Private Existence Algorithm. Even when we will write from now on only Binary Physics or only the Existence Algorithm, we intend Private Binary Physics or the Private Existence Algorithm.

It is important to note that not only does the Existence Algorithm affect the Private Binary Physics of each and every universe. In effect, any change of settings that does not contradict the axioms will establish a new universe. For example, if we change the initial state (see the discussion of the Fourth Axiom), or if we set a different response time for each cell (see the discussion of the Fifth Axiom), *etc.*

In this book, I will deal with the private physics of our Universe and the investigation of our own Private Existence Algorithm. A fascinating study of General Binary Physics and the many possibilities that are latent within it, I will leave for a later time.

Why was the specific Existence Algorithm and specific initial state that we see "chosen" for our Universe?

Of course, I have no answer to this question.

And I momentarily will digress from a physical discussion into a theological discussion. The very use of the word "chosen" already takes us out of the realm of physics...

But I could not avoid contemplating, that among countless combinations of Existence Algorithms and initial states—for our Universe in particular a combination was chosen that creates great complexity. It is easy to see that most of the combinations of Existence Algorithms and initial states create a universe possessing far less complexity.

The difference between a combination that creates complexity and one that does not create complexity, can be easily understood by observing the rules of two-dimensional cellular automata (or simply, the rules that determine how the next row will appear, based on the order of the white or black squares in the previous row).

On the one hand, there are many rules that create homogenous patterns or patterns that decay in terms of their level of change and complexity. For example, a rule according to which in the next interval each black square turns white and each white square remains white. This rule will produce negligible complexity after the first turn. Also a rule, which says that a black square turns white and a white square turns black, will create a pattern of minimal complexity.

On the other hand, there are a few rules that create the most complex patterns that superficially look random and unpredictable. An example of this can be seen in Rule 30 of the cellular automaton, which was presented by the Jewish physicist Stephen Wolfram in 1983, and also later in his book *A New Kind of Science*.

The truth table of this rule:

Initial Stat	111	110	101	100	011	010	001	000
Value of the middle cell in the next interval	0	0	0	1	1	1	1	0

According to this rule, if the initial state of the system is "Existence" surrounded on each side by "Placeholder", we will receive a most complicated pattern.

If we input the initial state in which it would be only "Placeholder", we will receive a pattern of much less complexity. In this example, we can learn that the initial state is no less important than the Existence Algorithm as a necessary condition for creating a complex universe.

Another example of the capabilities of cellular automata to create complexity can be seen in Rule 110 of the cellular automaton, which allows for complete programming by means of a cellular automaton (Turing complete). This means that it is possible to build any existing or possible computer program with this cellular automaton.

It is possible that the question why the Existence Algorithm and Initial State that produce such a great complexity was "chosen" for our Universe, is a circular question. It is easy to imagine a situation in which all possible General Physics combinations of existence algorithms and initial states have been formed, but only those that have a sufficient complexity have formed intelligent systems that are able to pose the question…

Chapter 6
The Private Existence Algorithm
(The Ten Rules of Physics)

We will get acquainted now with further two out of the ten rules of Private Binary Physics. Do not worry, by the end of the book, we will be familiar with all ten rules.

The second rule of Private Binary Physics – The Proximity Rule

> *"The Existence can move over only to one of the cells that encompass it directly."*

The significance of this rule highlights the impossibility of having a situation in which the Existence moves from one particular cell and appears at a distance of 10 cl away. We see this in observations. If we push an apple in Israel, it will move into the immediate space in the direction in which we pushed it, and will not appear suddenly in New York.

In General Binary Physics, it is understood that this rule is not compulsory. It is easy to imagine a universe in which an object appears a great distance away from where it has been originally Placed. As long as the location in which it is supposed to be is fixed and obedience to physical rules, there is nothing to prohibit the existence of such a universe.

The third rule of Private Binary Physics: The Rule of Non-mixing

> *"The Existence can pass over only to a cell with a Placeholder"*

The meaning of the rule is that there cannot be two Existences in the same cell. The derivative of this rule is that the quantity of the Existence cannot be reduced. The observations show us that if we push an apple into an apple that is standing next to it, it will move it and not merge with it. It is understood that in General Binary Physics, there is nothing to prohibit there being an Existence Algorithm that merges Existence, and thus in the universe that will be derived from it, the amount of matter can be reduced.

Chapter 7
The Fourth Axiom:
The Initial State

The Initial State is the condition in which the Existence and Placeholder were distributed in the universe, the instant before the Existence Algorithm started to run.

Contrary to the intuitive understanding that the Initial State defines the distribution of the Existence and Placeholder only in space, in fact it describes their distribution in the entire universe, including the depth dimensions. In most initial states, the Existence and Placeholder are distributed not only in spatial dimensions, but also in depth dimensions. In terms of Classical Physics, the Initial State determines not only the quantity of emptiness/matter in the first second of the Universe, but also shapes the entire future and past as it was at the instant that the Existence Algorithm started to run. It is a bit complicated, but in Binary Physics, since both the past and the future are represented by binary fields, they have an initial state, and they are constantly changing, just like the present. We will better understand this non-intuitive idea when we deal with the concept of depth.

The Existence Algorithm begins to define cell values only from the initial state onwards.

General Binary Physics allows for a lot of initial states, and in fact the only limitation on the quantity of initial states is the size of the universe.

The Initial State is an axiomatic definition. This is because it is not subject to the rules of the Existence Algorithm, and therefore cannot be defined within the system.

Of course, there may be particular situations, where even the initial state would meet requirements of the particular Existence Algorithm of that universe. In these situations, it would be possible to imagine through research the universe as it was before the initial state. However, here is a catch. It is true that we can imagine the universe as it was before the starting point, but for that very reason we would not be able to identify what was the initial state...

For example, it may be that our Universe was created a few seconds ago (or a few turns ago, if we talk in terms of Binary Physics). If it was created in such a manner that the Initial State, which was assigned a few seconds ago, is consistent with the rules of the Existence Algorithm that is operating at present, we would be mistaken to think that the Universe had been established much earlier. This is because we can recreate in our imagination how the Universe was before the initial state. Of course it would only be the fruit of our imagination, because it could be that before the Initial State nothing existed at all.

If we summarize, a consciousness, which is internal to the system and is confined by it, cannot know what was before the Initial State, and often it will find it difficult to determine the point of the Initial State itself at all.

The Fourth Axiom of General Binary Physics:

> *"There exists an initial state, in which and prior to which, the values of the cells are not subject to the Existence Algorithm."*

Just as there is a private Existence Algorithm for our Universe, similarly there is a **Private Initial State** for our Universe.

When we use the term "Private Initial State", we do this to distinguish our Universe. For simplicity, each time when we write "Initial State" only, we mean the Private Initial State of our Universe.

The different initial states affect the nature of reality in a very significant way. Universes, with the same Existence Algorithm, but with a different initial state, may appear different from one another beyond recognition. It could be that under the same Existence Algorithm, a highly complex universe would be formed under a particular initial state, whereas another initial state would trigger a decaying universe or an ordinary one, in terms of the complexity of the change in it.

In the framework of the study of Private Binary Physics, we will try not only to recreate our own Private Existence Algorithm, but also our own Private Initial State.

The observations of Classical Physics indicate that the Private Initial State of our Universe was what we call the "Big Bang".

In terms of Private Binary Physics, we describe the Initial State as the state with the greatest density of the "Existence" around the General Point of Reference (in space and in depth), whereas most of the Universe in areas distant from the General Point of Reference contained the Placeholder. It is important to emphasize that the Existence around the General Point of Reference was at high density not just in space, but also in depth, that is "time" was very dense at the point of the Big Bang, and not just space. We will see later that the density of the Existence at the point of the Big Bang in the plane of depth had a significant influence on the development of our private Universe.

Note: It is convenient to determine the General Point of Reference at the point of the Big Bang, due to the character of the Initial State of our private Universe, but this would not necessarily be the geometric General Point of Reference of our Universe.

Starting from this Initial State, the Existence started to spread out through the space of the Universe, according to the Existence Algorithm.

Chapter 8
The Fifth and Final Axiom:
The Queue

The observations indicate that things do not happen simultaneously. There exists a type of queue. One thing occurs after another.

The Third Axiom states that the change from the Existence to Placeholder is done in accordance with the Existence Algorithm. From observations of the Universe, we see that the change is performed in pulses. There is no drastic change that happens at once, but we rather see a small change that follows another small change, and so on. If we will liken this to a game, there is a queue. In each turn in the queue, a change is performed.

A large number of turns allows for a great number of steps, and a small number of turns allows for a small number of steps. A single turn allows for a single step.

The sense of time is formed from the relationship between the number of changes and the number of turns. The more changes are made in fewer turns, the more our consciousness interprets this as "time" that elapses more quickly.

Since we cannot prove the existence of the queue, it is determined by **the fifth and final axiom of General Binary Physics –**

> *"For every cell, there exists a queue, and in every queue, the value of the cell is determined anew, according to the Existence Algorithm".*

The queue is essentially the minimal "response time" of a cell in the Universe. In one turn, one change can occur in each cell, in accordance with the Existence Algorithm.

It may be that other consciousness, which exists outside of the system and is not subject to it, can grasp a simultaneous existence, in which all the changes take place at once. That is, an existence where the reaction time of the cell is zero and the speed of the system is infinite. It is reasonable to assume that minds, which are external to the system, can perceive reality without a queue and grasp the concept of infinity.

Our brain interprets the sequence of turns in the depth dimension as "time", just as it interprets the sequence of cells in space as an "image".

It is possible to stop time, but the queue always proceeds forward.

Time is a concept of consciousness, and therefore it can "stop moving". The queue is an objective physical concept, and therefore there is no possibility of stopping it. When consciousness does not see a change, it interprets this as if time stands still. But the queue continues to proceed forward, even if the clock is not moving…

The axiom of the queue in the framework of General Binary Physics allows for many interpretations of the queue, which are all valid.

For example, it is possible to imagine a universe in which each cell has a different response time. That is, the turn of one cell is equal to one-and-a-half turns of another cell.

It is possible to describe a universe with less variety, in which a group of cells have a different response time. In such a universe, you can throw a ball into a basket fast, but near the basket its motion will slow down.

One can describe a more variable universe, where not only different cells have different response times, but also the cell response time changes from one turn to the next. You would have no idea what will be the movement of the ball towards the basket. All of a sudden it speeds up and slows down. Suddenly a player next to you moves at a turtle's pace, and next moment the Superman is considered slow in comparison to him.

Which queue exists in our Universe?

The observations indicate that in the Private Binary Physics of our Universe, it is possible to determine a significant insight about the queue—it is uniform.

The meaning of "uniform" is, of course, that from the viewpoint of the observer

Internal to the system, every change in a single cell in the Universe always requires one turn. The reaction time of one cell is identical to the reaction time of any other cell in the Universe. In other words, there is a uniform queue in the system. All cells have an identical response time. Uniformity determines that the system is synchronous.

An observer inside the system, who is confined by it, can feel that different changes take different periods of time, but this is just an illusion that stems from the fact that our mind interprets several cells as one cell. We will discuss this illusion later in the chapter on the Resolution.

We have seen that in the framework of General Binary Physics, there are many possibilities besides a uniform queue that is the same for all cells. It is possible to say that the commonplace situation in a variety of possible worlds has a different response time for cells. That is to say, in the vast majority of the worlds, an asynchronous system operates. Our case, of a uniform response time, would be unusual and special.

There are three reasons that led me to the conclusion that our Universe is based on a synchronous system.

The first reason: simplicity.

In a situation of uniformity, it is possible to describe an Existence Algorithm in which every cell "accepts" the decision regarding changing its value in relation to the surrounding cells and in relation to its value in the previous turn—in an autonomous way. Synchronization of the system is what allows each cell to cooperate simply with all the surrounding cells without the need of external assistance. This creates a very simple mechanism of activation. The level of complexity of the system drops sharply, and you can set it in motion with a very simple Existence Algorithm. A synchronous system is also a system that is consistent with the First Rule of Binary Physics.

The principle of simplicity guides us in such a way that, without any evidence to the contrary, we prefer a simple interpretation.

The **second reason** why I came to the conclusion that our own private system is synchronous is because an asynchronous system contradicts the rules of Binary Physics that I have determined so far.

For example, the assumption of an asynchronous system negates the First Rule of Binary Physics, according to which the value of each cell is determined by the value of the surrounding cells in the preceding turn. If the concept of the queue is not uniform, then there is no meaning to this definition. In fact, in an asynchronous system, highly complicated calculations would be required to calculate the value of every cell in every queue due to necessity of scanning and calculating different turns of the cells in order to avoid contradictory situations. We will see that this computational complexity is almost insoluble, and perhaps is truly unsolvable. In any event, there is no reason to assume complex computation, and we will prefer a simple model, as long as the observations do not require us to do otherwise.

How can a particular cell, calculate its value based on the values of neighboring cells, if these values are uncertain? After all, if the response time of a neighboring cell is different, it could be that the data upon which our cell has relied is no longer relevant at the time of the change, since the turn of the surrounding cells is shorter…

The strongest argument in favor of a synchronous system is the observation of the speed of light as an absolute and objective measure.

If the response time of each cell is different, you cannot allow cells to "make" the decision autonomously because such a situation will cause chaos and many contradictions in the system.

Let us describe an initial state with the "Existence" in one cell. The "Existence" is said to move over, in accordance with the Existence Algorithm, to the neighboring cell, in which there is a "Placeholder".

In compliance with the Rules of Binary Physics 1–4, which I reviewed above, and because any change requires a turn, in the first turn after the initial state, the first cell goes from the Existence to Placeholder, and the second cell goes from the Placeholder to Existence.

Now let us suppose that the system is asynchronous. For example, let us assume that the second cell has a response time twice as long as that of the first cell. That is, the length of one turn of the second cell is equal to two turns of the first cell.

Now let us review the simulation. The initial state remains as it was. The turn of the first cell comes and it changes its value from the Existence to Placeholder. In the meantime, the second cell remains Placeholder. The Existence has disappeared.

This contradicts the Fourth Rule of Binary Physics—the Rule of Conservation of Existence. Also, we do not observe a situation in which we move an apple and it disappears completely for an instant and a few seconds later it reappears in the place, where we pushed it to…

Similarly, in an asynchronous system, another turn, external to the system and synchronizing changes between the cells with different response times, or, as stated before, an especially complicated Existence Algorithm would be required.

Fortunately, we have yet another strong proof for demonstration of the uniform response time in our Universe.

The speed of light reflects a wave of change among the cells. When a light beam moves, the cells change their values at the maximal possible speed. In other words, when we observe a light beam, we observe cells that are changing at their response speed.

> *The response speed = The inverse function of the response time = (response time)$^{-1}$*

If every cell had a different response time, we would not observe a light beam moving with a uniform speed in a vacuum, but rather changing its speed in different regions of space. I emphasize that the reference is made to a light ray that not changing its speed in a vacuum. Not to changes in the speed caused by the passage of a light ray through different materials, such as water or glass.

To explain this, let us turn to an example. Imagine a row of lamps.

We want to create the illusion of a "running lamp". We have to do the following: light up the first lamp in the row, in the next turn extinguish the first lamp and light up the second one, and so on. If the reaction times of all the lamps are identical, we will see the light "running" at a uniform speed. If some lamps have different reaction times, we will see the light "running" at variable speed along the row.

Thus, different response times of cells will cause the light speed to be not uniform, but rather variable. In such a situation, of course, there will be no meaning in observations of the light speed being the maximum possible speed.

From here, we get to one of the most important conclusions of Binary Physics.

The very fact that the maximum possible speed (the speed of light) is the same in every location in space indicates that the cells that comprise our Universe have uniform response times. And on the flip side, the limitation on the response time is what causes us to see the observations about the "speed of light", which we interpret as the maximum speed in the Universe.

If there was no uniform response time for every cell in our Universe, certainly there would be no ceiling to the highest possible speed. We would be able to accelerate forever.

Let us go back for a moment to the example with the row of lamps. Any DJ would tell you that you can run the lamps slowly. Then you can increase the running time more and more. But the limit of this speed is the response time of the lamps.

From the upper limit on the maximum possible speed in our Universe, we can learn one of the most important things in Binary Physics: the response time of the cells that makes up our Universe.

This insight will become the Fifth Rule of Private Binary Physics: the uniformity of the queue:

> *"The Queue is uniform for all the cells."*

In Private Binary Physics, since the queue is uniform and identical for all the cells—i.e., the response time of all the cells is identical—it is possible to measure the queues in uniform units. The units in which we measure the queue are called **"turns"** and marked **"tu"**. Each unit is one turn. **The turn is the response time of the cell.** The symbol for the units of the turn is 'tu'. That is, if we want to say that 3 turns have transpired in the system, we will say that 3tu have passed.

Hence the turn becomes the unit of measure of the dimension of depth, just as the cell is the unit of measure of space.

These two units are symmetrical and analogous. Since any change in space takes one turn, and since the depth is built of cells that are identical to space, it is possible to measure the depth in cells and space in turns. The use of the two different units is done only for clarity.

Chapter 9
The Dimension of Depth

Depth is one of two fundamental concepts of Binary Physics. The second concept is the Resolution, which I will deal with in Chapter 16.

> *Depth: "The dimensions of the queue"*

The concept of depth is a difficult one to grasp.

The reason is that we human systems are blind to the depth. We feel its existence, but do not succeed in seeing it.

Binary Physics certainly allows for the existence of sentient beings that see the depth or part of it, just as we see the dimensions of space.

We will see that this is only a matter of evolutionary progression of generations.

Depth is the father of the mental concept of "time", as we are familiar with it from Classical Physics: a second, a minute, an hour…

Because of our blindness, as human systems, to the dimensions of depth, we call our indistinct feeling with respect to them as "time". Depth is the physical reality, and our subjective interpretation of it is "time".

In Binary Physics, depth is a collection of physical dimensions from every aspect. Therefore, a particle in Binary Physics cannot advance on the depth axis if the particle in front of it is blocking it, just as a particle cannot move forward in space if another particle is blocking it.

In Classical Physics there is nothing that can prevent a particle from moving forward in time. Classical Physics treats time as a smooth stream that is clear of obstructions, while it is far from being so. An interesting and very significant consequence of this is that "the future" in Binary Physics influences the present no less, and sometimes even more, than the past. We will discuss this in more detail below.

When we say "time", we mean the comparison that we make regarding the speed at which various changes take place.

We compare the change of the location of a moving car to the change in the position of the clock hands. We take an arbitrary, subjective and varying rates of change, call it "a second", and decide that this is a measuring rod against which we will measure the rate of change of other events. Just as we take an arbitrary measure of length and call it a "meter", and compare other length measurements to it.

Both the second and the meter are arbitrary and relative standard measures. They vary with relation to the speed at which they move. A meter stick that moves quicker is shorter than a meter stick that moves slower, and a clock that moves faster measures a second slower than a clock that moves at lower speed. We will see later why exactly this happens.

It was convenient for us in the Middle Ages, because most human systems moved at similar and very slow speeds (of walking or riding) that (almost) did not affect the measurement.

As opposed to a meter, when we determine that ten cells in a cube of space are the standard measure of length, we receive an absolute benchmark of 10 cells. This standard measure will remain identical and not influenced by the illusion of speed at which our consciousness imagines that a certain system of Existence moves.

Thus, in the same manner, if we determine that 10 cells of depth are our standard reference measure of time, we will receive an absolute benchmark of 10 turns. One turn is the response time of the cell. Therefore, the highest speed at which the Existence can proceed in depth (in the language of classical physics in time) is one cell in one turn. The turn units, like the cell units, is not influenced by the illusion of speed, at which our consciousness imagines that a particular system of Existence moves.

Depth has three dimensions.

The Existence flows between the different regions through the dimensions of depth.

In fact, the space with which we are familiar (with its three dimensions) exists within the physical planes of depth.

Each dimension of depth is identical to the others, and all three of them are identical to the three spatial dimensions.

Each of the six dimensions of the Universe of Private Binary Physics (I remind you that two more dimensions belong to the realm of General Binary Physics) —the three spatial dimensions that we see and the three dimensions of depth to which we are blind to see— are constructed from identical cells, each containing the Existence or Placeholder. Incidentally, I will note that the seventh and eighth dimensions of General Binary Physics are identical to each of the other dimensions.

What distinguishes between one dimension and another is only the Existence Algorithm, which defines the manner of flow of the Existence between the cells.

The Existence Algorithm, that synchronizes the various directions of flow in each dimension, creates an illusion that the dimensions are different from one another.

Binary Physics separates between the relative-subjective-conscious units of time t and the units of depth, the units of the turn, which are absolute and unambiguous.

The time t is actually a comparison, that the observer who is internal to the system and confined by it, makes between one change and another in the system which he is observing.

The comparison between changes can be popularly exemplified by those in a quartz crystal (which is sometimes called a clock) or by changes in the location of a car. The subjectivity of the comparison stems from the interpretation of a viewer to the nature of the changes resulting from the "input device" (= the human system's mind), which is blind to depth and thus leaves unnoticed the objective source within the system that can serve for comparison.

If we look at both the clock and the car on the earth's surface, our mind is able only to compare one local phenomenon to another, and is unable to comprehend the fact that the Earth flies through space at a dizzying speed, and this has an impact both on the change in the crystal and the change in the car. Einstein established this principle as the Theory of Relativity, according to which the speed affects the time. The speed affects our subjective sense of time and the relativistic measurements that we perform. But it does not affect the depth—the objective queue. In Einstein's theory of relativity, there is no objective origin, which would serve for absolute comparison between the observed phenomena. This is depth.

There are two types of clocks in the system.

1. **Internal clocks.** These are clocks that are identical to the devices, which we are familiar with in our daily life. This can be, for example, a wristwatch that a human system wears in our Universe. These internal clocks are essentially a relative view into the three dimensions of the rate of change of one thing compared to another. Just as a wristwatch runs with the rate of change of a quartz crystal, which is based on a relative constant physical phenomenon and can be used for comparison with other changes in space.

2. Binary Physics requires another clock, one **external** to the system, which is worn by an observer external to and not confined by the system. This is the axiomatic clock of the queue.

The clock of the queue (or, by its more precise name, the dimension of depth) allows comparing objectively and unambiguously between the various changes in the system. The clock of the queue is not affected by speed. We will see later that the term "speed" is a conscious, illusionary concept that does not exist in the physical reality.

At time t, if an observer does not notice a change, time stands still from his perspective. This is unlike the dimension of depth $d(4)$, in which the turns continue to proceed, whether there is a change or not. The expression "time stands still" is true only for psychological and relative time t. For this time, there is no dimension. It exists only in the observer's mind.

The concept of depth along with the concept of resolution link between

Our subjective observations about relative "time", which slows down with increasing "speed", and the objective reality, in which

There is only one particle traveling at a constant "speed". I put the concepts of "speed" and "time" in quotation marks, because, as I mentioned before, these two concepts do not exist in the physical reality, but they serve as our mind's interpretation of the physical reality.

I sincerely hope that, after reading this book, it will be obvious how these concepts are formed in our mind and how their existence is consistent with the physical reality of one particle moving at a constant speed.

Chapter 10
The Fourth Dimension: The Dimension of the Queue
Indicated by "d(4)"

Depth is a physical dimension, which, due to the way the Existence movement through it, creates for us human systems the subjective feeling of time.

The numbering of the depth dimensions immediately continues numbering of the three spatial dimensions. Therefore, the first depth dimension is counted as the fourth dimension.

If the three spatial dimensions form a cube (as I mentioned in the axiom of the Binary Field, I use the concept of a cube only for convenience, which has no effect on the true manner in which the cells are arranged), then every unit in the dimension of the queue is a separate cube of space. I call space by the term "a cube of space", so that we can easily visualize it.

Let us explain this by means of an example. When an apple advances "in time" from 08:00 AM to 08:01 AM, what actually happens is that the Existence that makes up its pattern has passed from the spatial cube d(4)0 to the spatial cube d(4)1.

The spatial cube d(4)2 is what we call "the future" of the spatial cube d(4)1. Spatial cube d(4)0 is what we call "the past" of the spatial cube d(4)1, and so on...

The Fifth Rule of Private Binary Physics states that the queue is uniform for all the cells. Therefore, we can say that each progression in one spatial cube is a progression of one turn in the queue. Therefore, the units of the fourth dimension are units of turns (the cell's response time).

The above example, may be somewhat misleading, since it is clear that in every moment there is a huge amount of turns… and therefore it is more correct to say that the apple passed from cube d(4)1 to cube d(4) 80.5×10^{50}, since this is approximately the number of turns in a minute that is measured by a "static" clock at the Earth's surface. I will discuss later on at length how to calculate the value of a turn in relation to "time"… I would only like to mention that the units of "time" (the second, the minute) are relative units that depend on speed, whereas the turn is an absolute unit of measure. Therefore, the number of turns contained in a second or a minute changes with the speed at which the instrument measuring this time unit is traveling…

From observations in the Universe, we see that all change is made in the time domain. We do not see things moving only in space in "zero" time. That is to say, an apple that was on the right side does not appear suddenly on the left. Ostensibly, we could say that it is possible that the apple was not moving in "zero" time, but simply that its speed is infinite, and we only had a feeling that it was moving in "zero" time. But since we know from observations that the speed of our Universe is limited to the speed of light (and in the concepts of Binary Physics, is limited to the response time of the cell—the turn), this is not possible.

From here we can conclude that any movement in space requires a movement in depth. It is not possible to move only in space.

The existence of depth also seems to me as an obvious assumption.

Of the two possibilities:

(1) When I move in "time", the place from where I came, "the past", disappears the moment I leave it, or

(2) When I move in "time", the place from where I came, "the past", remains there, and the future into which I am about to pass, already exists for me to pass into it

I prefer the second possibility.

The assumption that depth exists, and the Existence just moves within it, definitely sounds more reasonable to me. As when I leave the living room to go to the bedroom, the living room does not disappear just because I left it. Similarly, when I leave one spatial cube (Time #1) and pass over to a second spatial cube (Time #2), the spatial cube which I left does not disappear.

If the fourth dimension did not exist, and the axiom of the queue would be negated, all the changes in space would be occurring simultaneously and at an infinite speed. This is something outside the limits of our imagination, and therefore outside the domain of physics.

Let me illustrate.

For simplicity, we will take a two-dimensional space (although exactly the same rules will apply to a three-dimensional space).

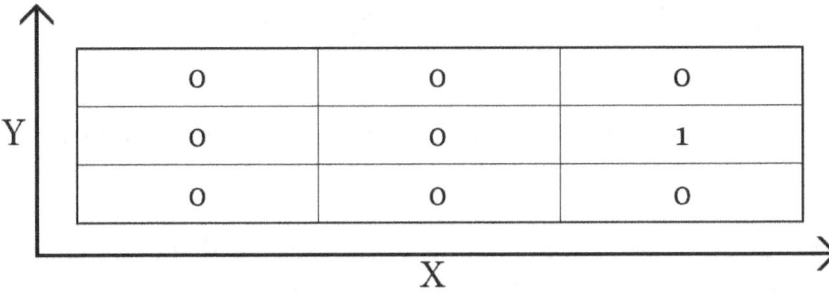

Note: In Binary Physics, the number in parentheses specifies the depth dimension we are dealing with.

In the initial state being described, there is one Existence surrounded by Placeholders. Let's assume that according to the Existence Algorithm this Existence must move 1 cell to the right. It cannot do this only in space. This Existence will have to proceed in depth as well. Any change to the value of the cell will require a turn in the queue, and every turn will require proceeding in the dimension of depth of the physical queue d(4).

Therefore, the result of this change will be seen only in d(4)1:

Y			
	0	0	0
	0	1	0
	0	0	0

X

Of course, according to the First Rule, the cell at d(4)0, or by its full name d(4)0 x2y2, will change to become Placeholder.

This change required one turn.

From the entire discussion above, the Sixth Rule of Private Binary Physics can be derived:

The Sixth Rule–Time (Symmetry of Space–Depth)

> "Every movement of an Existence in space requires it, Simultaneously, to move forward a cell in the depth dimension"

We saw that the derivative of the Fifth Axiom is that every change in space requires a change in depth. But where did we determine that the change must be "forward" in depth? (Forward in the sense of going from d(4)x to d(4)(x+1)) Why cannot the change be backwards in depth? Or sometimes forward and sometimes backward?

All the above options are valid in terms of General Binary Physics, and do not contradict the Fifth Axiom.

By the way, as long as the direction is uniform, there is no meaning to proceeding "forward" or "backward". These are terms that are simply more convenient for our minds to understand.

What possibility is relevant for our Private Binary Physics?

The observations indicate that in our Universe the Existence Algorithm defines a uniform and clear direction to proceeding in depth.

When I look at an apple, I see it proceeding with me in depth (what Classical Physics calls "time").

If we assume that there is no uniform direction to progression in depth, we would see states in which the apple would "freeze" in depth, or that it would move in a direction in depth opposite to my own. In such a situation, we should witness an observation in which the apple disappears. We do not have such an observation.

Since General Binary Physics allows for a universe where there is non-uniform progress in depth, so I can certainly imagine a universe where I am sitting and suddenly an apple appears next to me. Apparently in that universe, such an event would be something ordinary. I certainly would question in such a universe, whether the apple came to me from the next level of depth, or from the previous level of depth... I guess Private Binary Physics of that universe will search for the answer...

Is it possible that large systems that are built from a great amount of Existence move forward in depth, and very small systems made of a limited amount of Existence can also move backwards in depth?

The principle of simplicity leads me to choose uniform and simple rules as long as there is no need to do otherwise. There is no reason to assume that a system possessing a small amount of Existence will behave differently from a system with a large amount of Existence. In order for the Existence Algorithm to be as simple as possible, it must have uniform rules for all Existence. It seems that adding a rule requiring checking the amount of Existence in a system before deciding in which direction it will advance in depth in the event of change, would complicate the algorithm significantly. First of all, it is very difficult to define where one system ends and where a second system begins. Secondly, in such a situation, any movement will require the formula to perform a scan of the entire universe, and it is difficult for me to see how to continue with the principle of the cellular automaton in this situation.

In a situation of immobility in space, is there still required to be motion in depth?

The answer is yes.

Imagine that in a particular cube of space, there are my own system (I mean the system of my body) and the system of an apple.

Suppose that in the next turn my system moves "to the right". The apple's system remains static.

By the Sixth Rule presented above, movement in space requires movement in depth. Therefore, it is clear that the movement of my system will cause it to pass on to the subsequent level in depth.

If we assume that in a situation of immobility, the system of the apple does not proceed in depth, then from my perspective, when I proceed from the spatial cube $d(4)1$ to spatial cube $d(4)2$, the apple disappears!

Such a situation is not seen in reality.

From here we can derive the Seventh Rule of Private Binary Physics:

The Seventh Rule – The Rule of Trivial Movement in Depth:

> *"The default state of a static Existence in space, is to move in the next turn toward his direct cell in the depth dimension."*

If we apply the Seventh Rule to the above example, my Existence passes to the next spatial cube due to my movement "to the right", and I still see the apple, because it moves to the following spatial cube due to its trivial movement in depth. It is a trivial movement, because it is the default. There is no need for any "cause" for this movement to occur.

Can a human system see a single cube of space, even on the theoretical level?

The answer is no, because the human system is a complex system, any calculation requires a large quantity of turns. In order to grasp a spatial cube, the human system requires computing ability that is faster than the queue, and this is something that is not possible according to the Fifth Axiom.

Chapter 11
The Fifth Dimension:
The Dimension of the Depth Plane
Indicated as d(5)

This is the dimension of depth where all the spatial cubes of each queue takes place. This is essentially the plane that is received from the sequence of spatial cubes d(4) within one unit of d(5). It is understood that, because each spatial cube d(4) is itself constructed of 4 dimensions, it is a bit strange to call the dimension d(5) that contains them by the term "plane"… Certainly it is not a "plane" in the ordinary sense of the word. But it is convenient to imagine the entire sequence d(4) within a particular dimension d(5) as something spread over the surface of a flat plane, so I chose to use that word. Of course, there is no necessity that it will actually be the structure of the physical universe.

In d(5)0 exist all the "spatial cubes" d(4)0 up to d(4)e of the first queue.

In d(5)1 exist all the "spatial cubes" d(4)0 up to d(4)e of the second queue.

In d(5)2 exist all the "spatial cubes" d(4)0 up to d(4)e of the third queue.

And so on…

An observer in the dimension of the depth plane sees the three spatial dimensions + every dimension of the queue (the fourth dimension of depth) at the same time. That is, the observer of the fifth dimension sees from outside and simultaneously all the spatial cubes of the queue, one after the other.

He sees simultaneously the first spatial cube, the second, the third, *etc.* (in terms of Classical Physics, he sees the entire past, present and future at the same time, at each and every second… at every second, because the "past" and "future" are unceasingly changing, exactly like the "present").

This is a bit confusing, so I will give an example:

Try to imagine a movie director, studying a film (I hope that the concept "film" is still familiar to you, in the age of digital photography).

Each frame of the film is analogous to one spatial cube. Of course, the film is two-dimensional, but the principle is the same.

Each frame of film is one rectangular space of $d(4)$.

That is, the first frame is $d(4)0$, the second frame is $d(4)1$, and so on…

Suppose that this is a short film consisting of 24 frames.

A director, comparable to an observer in the fifth dimension, will be a director who is viewing all the frames simultaneously. He sees all 24 frames simultaneously.

What will the director see?

The answer to this question depends on when he will look at the film.

The first time he looks at the film, he will see it in the initial state.

In this state, the film is completely empty. All 24 frames are empty. This is d(5)0. The first frame of the film, while the film is completely empty, can be called d(5)0d(4)0, the second frame can be called d(5)0d(4)1, *etc.*

Now our director begins filming. Suppose he shoots one frame per second.

The first second that our director is filming is d(5)1. How will the spatial rectangles appear at d(5)1?

In the first frame d(5)1d(4)0 we will see the actress arriving at a pool. The second frame d(5)1d(4)1 will still be empty, as will all the frames up to d(5)1d(4)24.

The next second, which the director films is d(5)2. What will the spatial rectangles appear at d(5)2?

In the first frame d(5)2d(4)0, we will see the actress arriving at a pool. In the second frame d(5)2d(4)1 we will see the actress jumping into the pool. The third frame will still empty, as will be all the frames up to d(5)2d(4)24. And so on…

After 24 seconds that the director films, which is d(5)24, how will the spatial rectangles appear?

In the first frame d(5)24d(4)0, we will see the actress arriving at a pool. In the second frame d(5)24d(4)1 we will see the actress jumping into the pool. In the third frame d(5)24d(4)2 will see the actress swimming in the pool. And in each and every frame, we will continue to accompany the actress through the dramatic moment when she is drowning, to the happy end when the lifeguard saves her life… Something that will keep the viewers in suspense until the frame d(5)24d(4)24.

Past, Present and Future

The above concepts are relative.

To give a physical formula of what the future is like, I have to know what is my present spatial cube, for example, $d(4)1$. The next spatial cube $d(4)2$ is the "future" of the spatial cube $d(4)1$. This is in accordance with the Sixth and Seventh Rules of Private Binary Physics.

The spatial cube $d(4)0$ is the "past" of the spatial cube $d(4)1$.

But you cannot talk about concepts such at "the past" and "the future" without referring to the dimension of the depth plane $d(5)$.

Taking the above example of the director's film, what is "the future" of the spatial rectangle $d(4)1$?

It varies according to its location in the dimension of the depth plane.

In $d(5)0$— the initial state—the future of $d(4)1$ is a spatial rectangle $d(4)2$ with only a Placeholder in it (in our terms, the frame is empty).

In contrast, in $d(5)3$, the future of $d(4)1$ is a spatial rectangle $d(4)2$ in which the actress is swimming in the pool.

We see that every spatial cube in the fourth dimension has a different "future" in every unit of the fifth dimension.

But the question of what will be in the future is more complex.

When a human system asks what will be in the future, he wants to know what the physical appearance of the next spatial cube after the present one will be, **in the next turn** (and not how the future spatial cube appears in the current turn). that is to say, after some of the Existence in the current cube will have undergone a change in the next spatial cube.

I emphasize the phrase "in the next turn".

When a human system of the spatial cube d(5)0d(4)1 asks about the future, he asks about the spatial cube d(5)1d(4)2 in the following queue; he does not mean the spatial cube of the future in his current queue, d(5)0d(4)2.

Unlike the example with the film, which is static, the spatial cubes are not static and they change all the time.

There can be a situation (which is the common case) that:

In d(5)2 – in spatial cube d(4)1 the actress jumps into the pool in the "future" of this spatial cube, in d(4)2 the actress swims in the pool after passing of one turn.

In d(5)3 – in the spatial cube d(4)1, there is nothing, because all Existence of this spatial cube, which our actress is composed of, has passed to the spatial cube d(4)2, and in spatial cube d(4)2 we find the actress continuing to stand at the edge of the pool. She does not know this, but the reason that she has not jumped into the pool is because the Existence, which she is composed of in the previous turn, in the next spatial cube d(5)2d(4)2, blocked the possibility of her "present" Existence, in the previous turn, in the spatial cube d(5)2d(4)1, from passing and creating her system swimming in the water...Therefore her Existence simply moved to another place on the edge of the pool...

The "future" influences the present no less than does the "past"...

In this example, we see that the "past" has changed. The actress in d(5)3d(4)2 standing by the side of the pool remembers and thinks that also in her past she was still standing by the pool. But in the physical reality—if she returns to her physical past (the previous spatial cube in the fourth dimension) to d5(3)d(4)1, "she" (or more accurately, the Existence that composed her system) will already not be there. She (= all the Existence, which she is composed of) would have moved to d(5)3d(4)2...

In the above examples, I tried to illustrate the viewpoint of the observer in the fifth dimension, who sees all spatial cubes of d(4) at every unit (turn) of d(5).

How can it be that the actress is simultaneously in spatial cube d(4)1 and in spatial cube d(4)2? After all, it contradicts the Fourth Rule of Private Binary Physics—the Rule of Conservation of Existence?

To answer this question, we must first define—from a physical standpoint—what is the actress? After all, there is no such particle called "an actress"...

The "actress" is a pattern of the Existence in a particular spatial cube (later on we will see that the pattern of the actress is spread over a series of spatial cubes, and not over an isolated spatial cube... But, for the sake of simplicity, we will relate to her at present as a "flat" pattern in three dimensions only).

The Existence that comprises the pattern of the actress is not in the same place twice. In other words, there is a similar pattern of the Existence in the spatial dimensions in the two spatial cubes, but this is not the same Existence Particles. In each of the spatial cubes, different Existences make up the pattern of the "actress". When the Existence more or less preserves its pattern over the flow in depth, we imagine the actress existing in the past as well as in the present.

The more Existence that comprises a particular system advances further in depth, the more changes will happen in its pattern, until at a certain instant when it will not be possible to say that it retains its pattern anymore.

For example, the Existence that constitutes an apple at d(4)1 will also constitute it at d(4)2 without any change, or with minor changes. At d(4)1000 (again, the number is just for purposes of illustration), there is a chance that so many changes will have occurred in the pattern of the Existence, that we will see them as a rotten apple, and at d(4)10000 it could be that we will not be able to define the pattern of the Existence as an apple at all (maybe the Existence particles will be arranged in the meantime in the form of apple sauce?).

The Big Bang from the Viewpoint of Depth

An observer external to the system at d(5), who would observe our Universe, should see, according to the knowledge that exists today in Classical Physics, more or less the following situation:

At d(5)0, he would see the initial state of our Universe—an instant before the Big Bang. He would see the first spatial cubes of the dimension d(4) full of the Existence at a very great density around the General Point of Reference (the origin of the axes). As he goes further on and observes more spatial cubes, the quantity of the Existence and its density would decrease, such that the vast majority of spatial cubes would contain the Placeholder only, or a very limited amount of the Existence.

At d(5)1000000 (once again, 1000000 is for illustrative purposes), he would see the first spatial cubes d(4)0-5 almost exactly as they were at the instant of the Big Bang. This is because the Bang practically did not start with them, since they were blocked by next spatial cubes which had a higher density of the Existence. In spatial cubes d(4)5-1000, he would see the climax of the Big Bang, in the spatial cube d(4)10000 he would see the Universe more or less the way we recognize it today, and at the spatial cube d(4)1000000 he would see our Universe as it will be in our distant "future"—with Existence scattered and with almost no such systems as stars and planets... The spatial cube d(4)1001000 will be entirely Placeholder, since the Existence from the Big Bang still would not have reached it (assuming that all Existence at the Initial State of the Big Bang was concentrated, for example, in the spatial cube d(4)0–999...).

The spatial cube, which number is same as the number of the fifth dimension, is called **"the frontier of time"**.

> *Frontier of time: "The limit of the range of influence of one spatial cube on another."*

For example, if we are now in d(5)100, then the Existence that was in spatial cube d(4)0 can flow to (and influence) d(4)100 at most. The Existence from d(4)0 cannot at this stage reach the spatial cube d(4)200, because not enough turns have passed to allow it to do so.

At the initial state, when all the Existence is concentrated in the first spatial cubes (like in an initial state of the Big Bang type), the frontier of time was unique with respect to obstructions. It is the only layer which is unaffected by obstructions.

For us as observers who are blind to depth, it is very difficult to know whether we are in d(5)900000 or d(5)1000000. In both cases, our spatial cube would seem similar. But the spatial cubes of our "future" and "past" will appear different.

Of course, I have many clarifications for the above statements.

The first clarification: Contrary to the widespread view of Classical Physics—that the Big Bang happened in the distant past and has ended—there is no reason to assume this. It is possible that the Big Bang may still be occurring. It is happening in depth in the first spatial cubes, and the Existence that is found there still continues to proceed to the next spatial cubes, and continues to effect a major change to the Universe. We obviously do not see it because we are blind to the depth.

In my view, the likelihood that the Big Bang is still occurring in depth is greater than the assumption that it has finished. This is in light of the proportion between the size of the observed Universe and the distribution of the Existence in it. This relationship suggests to us the number of layers of spatial cubes that existed in depth at the Initial State. As the size of the Universe is greater on the one hand, and the distribution of the Existence in it is more uniform on the other hand—more layers of spatial blocks are required at the Initial State.

In other words, it may very well be that **the Big Bang continues to occur at the present time**. And probably the First World War as well. And the Holocaust. Of course they have changed from the way that we remember them. And it could be that they are being conducted in a different manner, but it is reasonable to assume that they exist as patterns of spatial cubes previous to ours. As we move backwards from the spatial cubes of our present, a more significant pattern is required to achieve certain likelihood of retaining some status in depth. Therefore, patterns like eating an apple will retain less time status than the pattern of World War I, which in turn is likely to retain less time status than the most significant pattern of the Big Bang, which, as we have noted, has a possibility of continuing to exist from the beginning of time (the Initial State).

The second clarification: I refer to the geometric center of the concentration of the Existence, at the instant of the Big Bang, as the origin of the axes, and I choose to place there the General Point of Reference. It is not necessary that the concentration of the Existence at the Initial State, which is called the "Big Bang", will be the geometric center of the Universe. But because of the nature of our private Initial State and the dominance of the Existence around a certain point, it appears to me to be correct to set the origin there, even if it does not constitute the geometric origin of the Universe.

The third clarification: It seems obvious to me, but nonetheless I will mention it. The "Big Bang" is an Initial State of our Private Physics, and a huge quantity of other initial states can be described, whether in a universe with an Existence Algorithm identical to ours, or in universes with other Existence Algorithms.

The fourth and especially important clarification: I pointed out that the observer in the fifth dimension sees in the initial state the first spatial **cubes** with the Existence at a very high density around a particular point.

Not necessarily the case I've used the plural arbitrarily: "cubes" and not "cube". Contrary to our intuitive comprehension that ignores depth, according to which all Existence was found at the time of the Big Bang at the maximal density in space, the Existence was actually at its maximum density in both space and depth. Already at the Initial State, at the time of the Big Bang, there existed what is referred to in Classical Physics as "time" for many years, and perhaps even millions of years. This is because the Existence of the Big Bang spread out over a huge quantity of spatial cubes at the Initial State—in a structure similar to layers of an onion—layer upon layer.

Since the Second Axiom forbids two Existence particles to be in the same cell, therefore at the instant of the Big Bang the "upper layers" of spatial cubes—the ones less "deep" (in relation to us) — first had to "empty themselves" and distribute their Existence, and only then the deeper caves started to empty out and flow. Therefore, I argue that the first spatial cubes—the deeper ones— still are filled with the Existence and continue the Big Bang… If we look at $d(4)0$ of our Universe, in our epoch, we will continue to see the maximum density around a particular point before the Big Bang. The "time" at $d(4)0$ is frozen such that this spatial cube is obstructed by all the following spatial cubes that have not emptied out as of yet…

I will give another example to help understand the nature of the Fifth Dimension.

I go from the bedroom through the corridor to the living room. Each transition from one place to another takes me one turn.

For this example, I am the "Existence" and the rooms are "Placeholders". In the initial state, there is only one "Existence". We will see later that it is easy to get confused and suddenly to create another "something out of nothing"…

Initial state—d(4)0—I am in the bedroom. The corridor is empty. The living room is empty.

First turn—d (4)1—The bedroom is empty. I am in the corridor. The living room is empty.

Second turn—d (4)2—The bedroom is empty. The corridor is empty. I am in the living room.

The above description can also easily be described in terms of Classical Physics if we imagine each turn to be a minute.

At 8:00 I am in the bedroom. At 8:01 I am in the corridor. At 8:02 I am in the living room.

But – this description is oversimplified and incomplete. It does not incorporate the dimension of depth.

Binary Physics states that any change requires progressing in depth. But depth was not created because of change, but is part of the Universe and exists from the outset as part of the Second Axiom.

That is to say, a more correct description of reality is required:

When I am in the bedroom at d (4)0, the spatial cube of d(4)1, which I am supposed to advance to, already exists! But when I was still in the bedroom at 8:00, the state at d(4)1 is that all the rooms are empty! Even the bedroom! At d4(1), in the first turn, all three rooms are empty. I have not reached any of them yet! This is the situation at d(4)2 as well.

When reading our initial description, it is possible to become confused and easily think that the Existence duplicated itself and I exist now simultaneously both in d(4)0 and in d(4)1. That is, in terms of Classical Physics, as if I have progressed in the future to 8:01, but I still exist in the past - in the bedroom at 8:00. This is something that contradicts the Fourth Rule of Private Binary Physics, the Rule of Conservation of Existence.

The correct and complete description, for which d(5), the dimension of the plane of depth, is required:

d(5)0 – Initial State
Spatial Cube—d(4)0—I am in the bedroom. The corridor is empty. The living room is empty.
Spatial Cube—d(4)1—The bedroom is empty. The corridor is empty. The living room is empty.
Spatial Cube—d(4)2—The bedroom is empty. The corridor is empty. The living room is empty.

d(5)1 – The depth plane of the first turn
Spatial Cube—d(4)0—The bedroom is empty. The corridor is empty. The living room is empty.
Spatial Cube—d(4)1—The bedroom is empty. I am in the corridor. The living room is empty.
Spatial Cube—d(4)2—The bedroom is empty. The corridor is empty. The living room is empty.

d(5)2 –The depth plane of the second turn
Spatial Cube—d(4)0—The bedroom is empty. The corridor is

empty. The living room is empty.
Spatial Cube—d(4)1—The bedroom is empty. The corridor is empty. The living room is empty.
Spatial Cube—d(4)2—The bedroom is empty. The corridor is empty. I am in the living room.

As can be seen in each of the planes of depth, all of the spatial blocks (the bedroom, the corridor and the living room) occur simultaneously, and I—the Existence—occur only once.

It is not intuitive for us as human systems, who are blind to the depth, that the spatial cube of the first turn transpires before the first turn occurs. The same can be said of the spatial cube of the second turn, and so on, inclusive.

In terms of classical physics, if my clock shows now 8:00, then the spatial cube of 7:59 continues to exist without me (because I have moved on to the spatial cube of 8:00).

Similarly the spatial cube of 8:01, which I am supposed to reach in the next minute, already exists—understandably without me. I shall get there in just a minute…

The understanding that all the depth dimension exists simultaneously and we are physically moving in depth just as we are moving in space, leads to several conclusions.

The first conclusion. "The past" and "the future" are changing just as present. It could be that a minute after I had gone to the spatial cube of 8:00 from the spatial cube of 7:59, Moshe went to the spatial cube of 7:59 from the spatial cube 7:58… That is—contrary to my intuitive thinking, according to which the past is just as I remember it— if I had come back to my physical past, I would have found Moshe there and not myself, but in any case he would have been different from what I remember and possibly very different…

Is it possible that I can go "back in time", that is, I can travel one turn back in the fourth dimension to the previous spatial cube, and I can find myself in a similar situation to that which I remember? The answer is affirmative.

The flow in depth is a wave-like movement. That is, it is most likely that the change from one spatial cube to another is not huge. That is to say, it is most likely that in the spatial cube that preceded mine, I would find Existence that forms the pattern that I call "Josef" (although it is reasonable to assume that there will be slight differences of the arrangement of the Existence in the second "Josef"). The pattern of "the other Josef" that I would find in the previous physical turn is built of another Existence than that of which I am composed. When my system of the Existence progresses to the next spatial cube (from d(4)3 to d(4)4 for example), the system of the "other Josef" advances from its previous spatial cube to my previous spatial cube (from d(4)2 to d(4)3 for example).

This situation is called a recurring event. Just like in a film of a movie, the difference between one frame and the other is not substantial, so the chances that the difference between one spatial cube and the next one are not great.

The greater number of turns I go back (that is, I go back depth, in "time"), the greater chances would be that the reality, which I find there, changed significantly. It is reasonable to assume that if I go back now to the turn that was two minutes ago, I would find a reality that is very close to that which I remember. But if I go back a great number of turns to the time of World War II, it could be that I would discover that the Holocaust did not occur at all...

The second conclusion. "Time"—depth—is composed of discrete cells, just as is space.

The third conclusion. "The Future" affects "the present" no less than does "the past". How can this be? According to the First Rule of Private Binary Physics, the value of each cell in the present turn is determined according to the values of the surrounding cells in the previous turn. When I write "surrounding", I definitely intend not only in space, but also in depth. That is, the value of a cell that is "before" the cell to which we refer to in depth (in the d(4) plane) influences it. For example, if there is "Existence" in the cell d4(0)z1y1x1, and there also is "Existence" in cell d4(1)z1y1x1 (i.e., the same cell, only in the following value of depth—I call this cell "directly following depth"), then it is clear that the Existence in the first cell cannot proceed to the directly following depth, because in the directly following depth, there already is an "Existence" (the Third Rule of Private Binary Physics—the Rule of Non-Merging—forbids an Existence from passing to a cell where there already is another Existence). In this way, "the future" influences "the present".

The effect of "the future" on "the present" is not negligible, but rather very significant, and it affects the causal connection no less than does space and "the past".

I will give an example. If there is a spatial cube of depth d(4)0, which contains an apple, and there also is an apple in the next spatial cube in depth d(4)1, exactly in the same location, then the apple of the first spatial cube cannot advance to its directly following depth. Its directly following depth is obstructed. What will happen is that the apple in depth d(4)0 that wants to advance to the next level of depth, will be forced to move from its place in space in order to advance! That is, what Classical Physics calls "the future of the apple" influences its direction of motion! When we will become familiar with the Tenth Rule, we will understand that the obstruction in depth that caused the apple to change its direction will be interpreted by us in space, as human systems blind to depth, as gravity!

When I feel that "I have decided" to move my hand to the right, it may well be that the reason for this "decision" of mine is that in my directly-following depth (which we call "the future") I moved my hand to the left, and therefore the depth on the left side is obstructed, and I am not able to move my hand to the left. Therefore, I have no choice but to move it this time only to the right (I am not referring my consciousness' illusion of "choice" that my mind creates for me).

The obstruction that we described above forms the only fundamental force of Binary Physics. The "Existence" will change its route because of only one reason, or one force if you wish, the obstruction, in either space or depth.

If, for example, in the initial state all the Existence was concentrated only in the spatial cube $d(4)0$, and all the other spatial cubes $d(4)1-e$ contained only Placeholders, then the flow of the Existence between the spatial cubes, under our private Existence Algorithm, would be uniform, and there would be no noticeable change in space in the dimensions of depth. The change will occur only if in the initial state, there will be a distribution of the Existence in depth as well, and thus the Existence of the "higher" spatial cubes will obstruct the Existence of the "lower" spatial cubes and cause a change in space. For example, a portion of the Existence in $d(4)0$ that is not obstructed in its next immediate level of depth by the Existence in $d(4)1$ will continue to flow without spatial change, whereas an Existence that is blocked will try to change its location in space, in order to overcome this obstruction, and thus create a change in space. Of course, the terms "higher spatial cube" and "lower spatial cube" are only to assist us in visualizing it, and there is no true concept of the "height" here. By a "lower spatial cube" I mean a deeper spatial cube, i.e. that is, a spatial cube possessing a lower number in dimension $d(4)$.

Because of its critical importance, I'll devote a separate chapter to the process of "obstruction", the fundamental force of Binary Physics.

The spatial cubes of each and every turn exists in parallel, and an observer in the fifth dimension views them simultaneously.

One more final example.

Suppose that every turn equals a minute on my relative clock.

An observer in the fifth dimension looks at $d(5)0$ and sees that when it is 6:00 for me (the spatial cube $d(4)10$), I break a vase. And when it is 6:05 for me (spatial cube $d(4)15$) I clean up the debris. I started at $d(4)10$ to allow for a "past", for the sake of the example.

The observer waits for one turn to pass. Now he looks at $d(5)1$ and sees that at 6:00 in the spatial cube $d(4)10$ I did not break the vase (for some reason that is concealed in the adjacent spatial cubes, "the past" $d(4)9$ or "the future" $d(4)11$. Since all the spatial cubes, i.e. all of my "past", "present" and "future" are his "present", he observes at 6:05 the spatial cube $d(4)15$, and sees me cleaning the vase. How is that?

The reason is that only in five more turns on the observer's clock in the fifth dimension the wave of time, in which I have not yet broken the vase, moves from the spatial cube $d(4)10$ to the spatial cube $d(4)15$, and then at 6:05 I will not clean up the vase…

Chapter 12
The Sixth Dimension:
The Dimension of Change in Depth
Indicated as d(6)

This is a dimension in which changes in d(5) occur.

In one unit of d(6), all the planes of depth of d(5)0−e transpire.

Let us try to put our ideas in order:

A unit in the dimension of the queue d(4) is one spatial cube.

A unit in the dimension of the plane of depth d(5) is a sequence of all the spatial cubes in one turn. It is possible to see in it both "the past" (the previous spatial cube) and "the future" (the next spatial cube) of every spatial cube.

A unit in the dimension of change in depth is a sequence of spatial cubes over the sequence of queues. In fact, one unit in the dimension of change in depth contains the entire existence of the Universe throughout all "time", from the Initial State until the end of "time".

An observer in the sixth dimension simultaneously sees all that was and will be, and witnesses the entire process of change and how it comes about at the same time. A clarification: Concepts "that was" and "that will be" do not exist for the observer in the sixth dimension, because all our "times" constitute his "present", they are all one unit of his time.

An observer external to the system that studies one unit in the dimension d(5) sees a static universe. He sees the entire sequence of the spatial cube in their

Current order.

At the instant that a turn transpires and a change is created, and we pass on a unit in d(5), a sixth dimension is formed—a dimension of change in depth—the dimension that allows change and movement.

When a turn transpires, and as a result of this, the values of the cells are reassigned, and the Existence flows in depth, and consequently all the spatial cubes that comprise e.g. d(5)10 change, then dimension d(6) is formed.

An observer external to the system who studies the sixth dimension d(6) simultaneously sees all the spatial cubes, changing at every moment, and observes the flow of the Existence into them.

An observer in the sixth dimension sees, at every instant, all of space in all its depth, and the change that transpires in it.

An observer in the sixth dimension sees in every turn "the past", "the present", and "the future" of every spatial cube, and when another turn passes, he sees how all the times—"the past", "the present", and "the future"—are changing.

See the diagram of the spatial cube d(6) at the end of this chapter.

Binary Physics determines that the entire "time plane" is not static. What we call "the past" and "the future" is changing, just as does "the present".

If my "present" is the spatial cube d(4)1000, then when a turn passes and the values of the cells are determined anew, then the new "present" of the system of the Existence that contains my "consciousness" will be d(4)1001. In this situation, the spatial cube of my physical "past" will be d(4)1000. But when the Existence that comprises my system passes from d(4)1000 to d(4)1001, the Existence from the spatial cube d(4)999 passes to spatial cube d(4)1000, and changes the spatial pattern of the Existence there. That is, if I go back to my physical past, I will find it different from what I remember.

The same is valid for the future.

If my "present" is spatial cube d(4)1000 and my physical "future" is the spatial cube d(4)1001, then when one turn transpires and the Existence from my present spatial cube passes to the next spatial cube, they will influence and change the future, since also from the outset some Existence in the spatial cube d(4)1001 will obstruct the progress of the Existence from my present spatial cube and thus cause them to move in space in their passage from the spatial cube 1000 to spatial cube 1001. Therefore, when an observer in the sixth plane contemplates d(5)1000 and d(5)1001, he will see two cubes of d(4)1001 that are different in their pattern of the Existence.

I will try to illustrate the difference between the dimensions by an example of a movie film.

A unit in dimension d(4)—the dimension of the queue—is analogous to one frame in the film.

A unit in dimension d(5)—the dimension of the depth plane—is analogous to the entire film, in one queue.

A unit in dimension d(6)—the dimension of the change in depth—is analogous to the entire film, the length of the entire process of change—at first we see it as empty, afterwards when only one frame is photographed, then when two frames are photographed and so on…

Of course we as human systems that are blind to depth will say that these are not "a lot" of films, just one film that is changing… Our brain compresses all the "information" of d(6) and interprets it as a change…An observer external to the system in the dimension d(6) will see physically many planes of the spatial cube, where each one represents another film…

The film example creates another distortion in gaining a true understanding of d(6). The film is a static object, in the sense that from the instant that we have taken a picture on Frame #1, this picture is fixed and does not change. On the other hand, in dimension d(6), during the passing of each turn, all of the frames are continuing to change. The Existence from Frame #2 tries to pass to Frame #3 (those that manage to overcome the obstruction of the Existence that is found in Frame #3 also succeed in passing). The Existence from Frame #1 tries to pass to Frame #2, and so on. Therefore, an observer in the Sixth Dimension will see all the spatial cubes in all the planes of depth changing at every turn. In our terms, he will see like us the "present" changing, but at the same time he will see also our "past" and "future" changing. From the view point of an observer in the sixth dimension, all times—past, present, future and more (the past that was and the past that will be, the future that was and the future that will be) are all one.

In a single turn, the observer in the Sixth Dimension sees all the time planes of d(5) sequentially, i.e. he watches a "movie" of the entire sequence of "physical time". An observer in the Fifth Dimension "is blind" to the way in which the spatial cubes appear before the turn transpires.

	An observer in one turn in Dimension: d(4)–The Dimension of Depth in the Physical Queue–Linear Space						
	Living Room	Corridor		Kitchen			
One frame of the Movie	1	0		0 d4(2)			

	An observer in one turn in Dimension: d(5)–The Dimension of the Depth Plane					
	Living Room	Corridor	Kitchen			
	0	0	0	d4(0)	d5(2)	
All of the frames in the movie in one turn	0	0	0	d4(1)		
	0	0	1	d4(2)		

	An observer in one turn in Dimension: d(6)–The Dimension of Change in Depth						
	Living Room	Corridor	Kitchen				
	0	0	1	d4(0)	d5(0)	d6(2)	
	0	0	0	d4(1)			
	0	0	0	d4(2)			
	Living Room	Corridor	Kitchen				
Filmstrip that changes over the course of filming	0	0	0	d4(0)	d5(1)		
	0	1	0	d4(1)			
	0	0	0	d4(2)			
	Living Room	Corridor	Kitchen				
	0	0	0	d4(0)	d5(2)		
	0	0	0	d4(1)			
	1	0	0	d4(2)			

	An observer in one turn in Dimension: d(7)–The Dimension of the History of Depth						
	Living Room	Corridor	Kitchen				
	0	0	1	d4(0)	d5(0)	d6(0)	d(7)0
	0	0	0	d4(1)			
	0	0	0	d4(2)			
	Living Room	Corridor	Kitchen				
	0	0	0	d4(0)	d5(1)		
	0	0	0	d4(1)			
	0	0	0	d4(2)			
	Living Room	Corridor	Kitchen				
	0	0	0	d4(0)	d5(2)		
The entire "filmstrip" as it is in each and every turn	0	0	0	d4(1)			
	0	0	0	d4(2)			
	Living Room	Corridor	Kitchen				
	0	0	1	d4(0)	d5(0)	d(6)1	
	0	0	0	d4(1)			
	0	0	0	d4(2)			
	Living Room	Corridor	Kitchen				
	0	0	0	d4(0)	d5(1)		
	0	1	0	d4(1)			
	0	0	0	d4(2)			

Figure 1: Sketch of the Spatial Cube of d(6)

Dimension of representation

The dimensions d(5) and d(4) are physical parameters of the Existence. d(6), being distinct from d(5) and d(4), is a dimension of representation. The Existence in d(6) does not exist in the usual sense. It is a conceptual Existence. Therefore, in d(6), the Fourth Rule of Private Binary Physics—the Rule of Conservation of

Existence—is inapplicable. Thus, in one unit of dimension d (6), we will see the same Existence in a number of planes of depth. The intention is that in the dimension d(5) we will see the physical Existence in a spatial cube d(5)0d(4)1 passing to spatial cube d(5)1d(4)2, whereas in dimension d(6) we will see that Existence in parallel, also in the spatial cube d(5)0d(4)1 as well as in the spatial cube d(5)1d(4)2. That is, we will see the same Existence twice. Therefore d(6) is considered the dimension of representation.

The consideration in the decision to establish d(6) as the dimension of representation, and not of Existence, has aimed to present the information in a more convenient way, and in any event, all the dimensions are a fiction of our consciousness. They are a way in which our consciousness arranges the information. If d(6) were also a dimension of existence, it would be very cumbersome to represent the development of movement in d(5) in this dimension.

Chapter 13
The Seventh Dimension: The Dimension of the definitions of the initial state
Indicated by p(7)

Each unit in the seventh dimension represents a new initial state and contains whole private physics of a particular universe, from the dawn of time (the initial state) until the end of time (the final state).

The entire Seventh Dimension, p(7)0–e, already is in the realm of General Binary Physics.

Both the Seventh and Eighth Dimensions are called dimensions of definitions: The Seventh Dimension defines the initial state of the universe, and in the next chapter, we will see that the Eighth Dimension defines the Existence Algorithm that operates in the universe.

One unit in the Seventh Dimension contains all the dimensions of change in depth d(6)0–e. That is, it contains the entire history of change of the Existence, from the time of the Initial State, for the entire length of the plane of depth, for each and every turn.

The seventh Dimension contains all alternative histories that may be from all possible initial states, under the same Existence Algorithm. In other words, we will find in it not only the whole history of the "Past, Present and Future" in each and every queue of turns in our Universe, but rather histories of many universes, which are totally different from ours in the distribution of the Existence, but very similar in having the identical Existence Algorithm.

For example, if our Universe is $p(7)0$, then in universe $p(7)1$ there will be a different initial state. At the initial state of $p(7)1$, we will perhaps find a uniform distribution of the Existence throughout the length of the entire space-time (space-depth). In such a situation, it is reasonable to assume that we will not find at $p(7)1$ galaxies that cluster around black holes, but rather a uniform and remote distribution of individual patches of the Existence. That is—a universe without stars or planets.

The word "history" is a bit misleading in the context of the dimension $p(7)$, because of its allusion to "the past", whereas here I refer to history in the sense of information and retention of information that represents the entire spectrum of change that we call "time", viz. "past, present, future ".

Going back to my favorite example of the film, if the Sixth Dimension is a lot of films of the same movie that keep changing, then the Seventh Dimension is Cinema City. It contains a lot of movie theaters, where a different film is being shown in each one – another dimension $d(6)$ in parallel.

The difference between each film in adjacent theaters are minor. For example, in the first theater $p(7)0$ Superman's suit is blue, and in theater $p(7)1$, Superman's suit is red. In theater $p(7)1000$, the changes are already so great that it is no longer the movie Superman, but rather Iron Man, and in theater $p(7)10000$ they are showing the movie "Titanic"…

In summary, in all "movie theaters" of dimension $p(7)$ all movies that you can imagine—each representing another initial state under our private physical rules —are being shown..

Chapter 14
The Eighth Dimension:
The Dimension of the definitions of the Existence Algorithm
Indicated by p(8)

Each unit in the Eighth Dimension represents a different Existence Algorithm. Within one unit of the Eighth Dimension, every unit of the Seventh Dimension displays the same Existence Algorithm, with a different initial state.

One unit in the Eighth Dimension displays the entire dimension p(7)0-e—all possible histories of all spatial cubes throughout the entire depth plane in each and every turn—under a specific Existence Algorithm.

The entire Eighth Dimension, p(8)0-e, like its partner, the Seventh Dimension, belongs to General Binary Physics.

In fact, every combination between a particular unit of the Eighth Dimension (= a specific Existence Algorithm) with a particular unit of the Seventh Dimension (= a specific initial state) creates a new universe with a Private Binary Physics of its own.

The Eighth Dimension incorporates all different options for arrangement of the Existence and all possible regularity (algorithm) of his movement. Therefore, all that was and will be in our or in any other universe—as long as it is possible and is found to be within the framework of axioms—will be within the framework of the Eighth Dimension. All General Binary Physics is folded up into the Eighth Dimension.

By means of the eight indicators, Binary Physics can give an unambiguous definition for everything that was and that will be.

And also for everything that never was and never will be, except in our imagination.

Alongside the enormous wealth of descriptive ability of the Eighth Dimension, it is also the ultimate limit of human imagination. To imagine things beyond the Eighth Dimension would require an external observer who is not confined by the system.

The entire Eighth Dimension p(8)0-e is a single unit in the Ninth Dimension. An observer in the Ninth Dimension sees all the worlds, at all times, in all possible types of physics—simultaneously.

What is the second unit of the Ninth Dimension, and the third and fourth? This knowledge goes beyond the boundaries of the axioms, and therefore is reserved for beings more exalted than us. Certainly more exalted than me.

Chapter 15
Resolution

We have reached the second fundamental concept of Binary Physics. The first was the concept of depth.

> *Resolution – "Imaginary reference that gives an interpretation to a number of cells as if they were one cell, which value is a superposition of the values of all its component cells."*

If Binary Physics argues that there is only one particle, how does human mind see a composite image that appears to be made of dozens of different forms and materials?

The resolution is the bridge that links the observations, which our consciousness sees in space and depth, and the physical existence of the constant universe—the fundamental resolution—the Binary Field itself.

Any observation at low resolution is an imaginary picture, which is based on a mere reflection of the physical reality of the fundamental resolution.

As we know, when we watch television, we do not really see a picture. We see a tremendous number of pixels. Our mind as possessing one consciousness is able to grasp only one reality. Therefore, it shrinks the reality comprised of tens of thousands of pixels into a single image.

We have already seen that if something within the system operates in a particular way, it is not accidental, but it is due to the nature of the whole system, and therefore each event is indicative of the entirety.

We can therefore infer that the way in which we view a television is an indication of the general way, in which our mind perceives the physical reality that we live in.

When we see an apple, we actually see a system of a great amount of Existence, which our consciousness has compressed and presented to us as a single picture.

Our minds are actually limited to looking at an apple at low resolution.

A more sophisticated consciousness will be able to look at an apple at a higher resolution, and to see the Existence from which it is composed.

The resolution is not a physical concept, but a concept of consciousness. It is not possible to explore the reality that surrounds us without referring directly to the instrument by which the exploration was conducted, i.e. the consciousness of human systems and its limited capabilities. Since the consciousness of the human system is built in such a way that it observes physical reality at a low resolution, while it presents a lot of data as a single datum to reconstruct the physical reality, we have to do "reverse engineering" and extract out the high resolution of the low resolution in which the observations have been made.

The resolution is indicated by the English letter R. The value of the resolution will appear in parentheses.

R (0) – the fundamental resolution – the constant universe – the binary field itself – is the highest level of resolution. At this level of resolution, we see the physical universe as it really is. We see each and every cell individually. We see every Existence and Placeholder. Therefore we call this resolution "the fundamental resolution". At times this resolution is referred to as the constant universe. At this resolution, there is no randomness or uncertainty. There are no variables, only constants.

You can imagine us human systems to be like people who are sailing on a ship in an uncharted ocean. From time to time we come upon the edge of the mountain that stands out. According to the outline of the mountains, the captain must map out the ocean floor. The tops of the mountains are the mathematical constants that remain at low resolution in which we view reality: π, the gravitational constant, Planck constant *etc*.

The ocean floor is the maximum resolution – the constant universe. All the parameters there are mathematical constants.

At the highest resolution, all measured data are "mathematical constants." The more the resolution decreases, the more the constants are blurred and appear to be relative indices. At our resolution R(h), relatively few "constants" remain. These constants are our hints by which we can reverse engineer the system of the constant universe. By means of these constants, we can reconstruct the Existence Algorithm.

If we want to understand how all the mathematical constants that exist in the Universe came about, we must seek out answers in the geometric structure of the cells and the Existence Algorithm as they operate in the constant Universe—at the fundamental resolution.

All rules of physics must be derived from the constant universe – the fundamental resolution. If an observation cannot be derived from the constant universe, we have explained it wrong.

R (1) – One resolution over the fundamental resolution. In this resolution, all 81 cells will appear to the observer as one cell.

In our own Private Physics, each spatial cube consists of 27 cells. Since the contraction of resolution has been performed simultaneously in both space and depth, in essence the decrease of resolution causes a conscious "connection" of 3 spatial blocks: the current one, the one before, and the one after. In this way, we come to 81 (=27*3) cells.

In any increase in resolution, consciousness "contracts" all 81 cells and turns them into a single imaginary cell.

The formula in our Private Physics is 81^R.

It is important to note that we are not speaking of merging or connecting cells. The resolution only presents a conscious idea, which exists only in the mind of an observer. The resolution explains to us how the observer interprets the objective physical reality.

R (2) – At Resolution 2 all 6561 (=81^2) cells appear to an observer as a single cell.

R (3–e) – At Resolution 3 all 531,441 (=81^3) cells appear to an observer as a single cell. Thus the value of the resolution reflects the number of cells that are contracted in the observer's mind into a single perceived unit. The value 'e' reflects the largest number that exists in the Universe. It is possible to say that in this value the entire Universe appears as one unit. Sounds illogical? Look up and see how you perceive an entire galaxy as a single point in the night sky.

Note: The numerical value of the resolution is a bit confusing; this is because the higher its number, the lower the actual resolution.

Resolution R	Number of cells=81^R
0	1
1	81
2	6561
3	531,441
4	43,046,721
5	3,486,784,401

I have not succeeded yet to prove it, but Binary Physics hints at the fact that the prime numbers result from the resolution. That is, these are the fundamental numbers into which it is possible to "contract" the cells and increase or decrease in resolution. In order to understand how the prime numbers came about, we have to understand better the way, in which our minds interpret resolution.

R(h) – the resolution of human systems – the observed Universe. This is a very special resolution. Its specialness is not objective but subjective. This is the resolution in which our consciousness, as human systems, observes the Universe. Since currently all physical research is performed by the minds of human systems, it is clear that the way in which they interpret the Universe is critical to understanding the objective physical reality. The gap between R(h), the resolution of human systems, and R(0), the constant resolution of the Universe, is the key and the bridge to understanding the physical reality, which we live in.

Just as there is a difference in many areas between one human system and another, so there is a difference between the resolution at which the mind of one human system observes the Universe and the resolution at which the mind of another human system observes it. The observations indicate that the difference is not great. Even in the Guinness Book of World Records, we are not familiar with people who are able to see rocks on Mars without a telescope or bacteria without a microscope. Therefore, it is possible to attribute small differences in the resolution of human systems, as one representative resolution.

It is important to emphasize that our consciousness has developed in an evolutionary manner to display to us the Universe at a specific resolution, and human systems do not have the possibility of perceiving the Universe at a different resolution. When we want to look at things at a different resolution, we require aids to translate the observations for us to our resolution. Popular aids are things such as the microscope or telescope. For example, even when I use a microscope, I do not see the germ at its true resolution, but rather I see the microscope's translation into my resolution.

The fact that we do not have observations of humans who see matter and light in a completely different manner, reinforces the belief that human systems observe the Universe at the same resolution.

In this context, it is worth mentioning that since the resolution acts simultaneously in both space and depth, if a human system sees space at a resolution different from $R(h)$, then inevitably he will also see depth at another resolution. This means that at a particular given speed, his seconds will be composed of a different number of turns in depth than one second of another human system. When you will read the chapter on the Birth of Materials, you should understand that this would cause one person to see all matter in a manner totally different from how we see it.

The understanding that resolution is a concept that speaks about depth, no less than space, is very important. Only with low resolution of depth, our conscious can compress the depth, and enable the interpretation of different patterns of Existence in depth, as different materials with different properties. After all, the entire difference between the particles in space is due to their different pattern in depth. Without low resolution, there could not be differences between the patterns in depth, because there is no depth to attribute. At the fundamental resolution, all is "flat", and there is only one form of "matter": the Existence.

Later on in the book I present the calculation of the value of the resolution R(h).

An imaginary cell – an imaginary cell is an observation at low resolution of a number of cells that are compressed in the observer's mind and are seen as one cell. For example, at R(1) an imaginary cell will contain 81 physical cells of the Universe. This concept is sometimes referred to as "imaginary Existence". This is because the brain of human system is blind to cells and sees only what they contain, the Existence. Then we can say that the superposition of all Existence together creates one imaginary Existence at low resolution.

How is the location of the Existence calculated at a lower resolution?

When most of the Existence has passed to another location, consciousness relates to this as if the system has moved.

Consciousness is blind to an isolated Existence, and sees all Existence as a kind of average. Therefore, reference should be made to most of the Existence. When you look at higher resolutions R(1), R(2), *etc.*, it is difficult to see this average. It is easier to understand this when we look at lower resolutions such as R(h), which is the resolution at which $81^{1.343*10^{50}}$ cells are considered a single cell.

The Resolution of Depth

Since we are blind to the depth, when we talk about resolution, we are imagining the compression of spatial cells into one imaginary cell at the resolution of human systems.

It is difficult to imagine that our view of the Universe at a low resolution affects not only space, but also depth, which we call time.

In other words, when we look at the Universe, we see not only space at low resolution, but also the depth ("time") at low resolution.

Because the Universe is symmetrical and depth is identical to space in terms of its cellular structure, it is perfectly clear that if the resolution affects how we compress the cells in space, it also performs a compression of cells in depth.

At human resolution, we see any number of cells of depth as one imagined cell of depth.

In terms of Classical Physics, we see any number of "true" time units as a single unit. Our brain actually shrinks "time" to create a single image, just as it shrinks space.

In terms of Binary Physics, we see a huge sequence of cells along depth—as the present—as the unit of time consciousness—such as the second, for example.

I chose the second as the basic psychological time unit of the human consciousness, because it is more or less the basic time unit, which the majority of humanity can grasp as the "present".

If resolution would only operate in space, then we would see observations and a Universe with only one type of matter.

Of course, we could determine that this matter is not "uniform" as our brain presents it to us, but rather consists of small fundamental particles.

However, the compression of the Existence in space is not enough to explain our observations of different forms of Existence at points in space. On the one hand, we see the existence of light rays, while on the other hand we see electrons and protons.

The observations show us that even in space itself, the essence of a light ray is fundamentally different from the essence of the proton. Even within electromagnetic rays, gamma radiation is different from ultraviolet light, which are different from infrared rays.

I will explain this by an example.

On a television screen with two modes for each pixel—black and white—we can receive pictures only in black-and-white (or their combinations that create the illusion of gray), whatever might be the structure of the pixels on the screen.

In order to obtain a color image on the screen, we need pixels of red, green and blue (in addition to the "off" position, which creates black).

In Binary Physics, there is no possibility of suddenly combining additional particles besides the Existence and Placeholder. Thus the situation is similar to a black-and-white TV screen.

How the illusion of different colors and different materials is created? It's the result of the resolution of depth.

To understand how all these materials have been created, we have to understand how the resolution in depth influences our consciousness. This is treated in the chapter on the Birth of Materials and the Nature of Mass.

Why has the mind been developed that sees space and depth at a low resolution, and essentially collapses a lot of information into a kind of single imaginary unit?

The limited computing resources of the mind on the one hand, and the limited means of input means on the other, caused the mind to evolve in such a way, that only a mind that has succeeded—despite its limited ability to interpret the partial information in such a way to preserve the system that contained it—survived.

Since the consciousness itself requires a complex system which Based on a vast amount of interactions of the Existence, consciousness can develop only at low resolution. A system that exists at low resolution interacts with other systems that are at a low resolution, and therefore the detection of other systems at low resolution is what is most relevant to it.

A consciousness that has managed to display the image of reality at low resolution—in a way that practically helps to preserve its own system— survives better than consciousness that sets upon another balance and utilizes its limited computing resources for the benefit of perceiving reality at a higher resolution (i.e., gaining a truer picture), and pays a price for this by processing a lot of information that is not required to preserve its own system.

Human systems do not see reality as it is; rather their minds imagine the visual input data in a way most relevant for its survival. We see an apple, not because it exists, but rather because to imagine an enormous number of Existences as one system, that is called an apple, helps our system survive. When we want to eat an apple, we pluck it as one entity. Therefore, a human system has no survivalist reason to invest energy in seeing the system of the apple at a higher resolution.

The ability of the mind to compress reality in both space and depth into a simple imaginary picture, which helps preserve its system, has been victorious in the evolutionary battle.

It should be noted that evolution definitely can provide for a consciousness that can extricate us from blindness to depth, and perceive it. It is possible to describe such a consciousness in both General Binary Physics and Private Binary Physics.

I certainly could imagine conscious systems that look at the expanse of depth (time), just as we look at space.

It is obvious that this property, if it would start to develop, would begin from the most basic capability of perceiving depth, perhaps in a way that is reminiscent of the difference between the first systems that managed to detect light, as compared to the complex system of the eye that is familiar to us.

Can there be a consciousness internal to the system, which will be able to see the Universe at the fundamental resolution?

The answer is: No. In order to see the "Existence", there must be a system with a reaction time that is faster than the maximum reaction time of the cell; this contradicts the Fifth Axiom—The Queue.

Chapter 16
The Obstruction
The Fundamental Force

The definition of Force in Binary Physics:

Force = "Preventing change – The interpretation of the mind to an obstruction that is external to the system, either in space or in depth."

Obstruction - "A situation in which an Existence can not move to another cell because that cell already contains an Existence."

In a sense, this is a definition opposite to that of Classical Physics, which is, in general: Force = "causing a change". Therefore, a small value of "Obstruction" in Binary Physics equals a large value of "Force" in Classical Physics. These are two values that constitute a mirror image of each other. Despite this technical difference—Force and the Obstruction are two ways of expressing a single physical entity.

The only physical observation that exists in the Universe is the observation of change (aside from a static observation of the very presence of the Existence and Placeholder).

The change is reflected in the "movement" of the Existence from cell to cell.

We will see in Chapter 19 that change is what Classical Physics calls energy.

If change is the only existing observation, then we must ask: What is the force that can affect change?

In Private Binary Physics, there is only one force that can affect change and even prevent change – an obstruction.

An obstruction has the ability to stop or slow down change.

In Classical Physics, when I perform work, for example, by pushing a cart, I exert a force on it over a distance and give it energy.

In Binary Physics when I do work, I remove an obstruction in depth over a distance, and thus allow the system of the cart to slow down less, and actually proceed forward at a faster speed, up to the maximum speed, which is derivative from the response time of the cell. When I remove the obstruction, I allow increasing the rate of change of the wagon's system. I will explain many more of the concepts that I wrote in this paragraph later on. But that is the principle.

If there were, for example, various types of energy in our Universe (which is not consistent with observations), then there would be reason to search for a number of forces, where each force influences a different type of energy.

In Classical Physics without exerting a force, a body will not change its speed.

In fact, it could be said that the concept of force in Binary Physics is opposite to the concept of force in Classical Physics. In Binary Physics, the basic movement is the maximum possible speed (the speed of the response time of the cell), and force slows it down or stops it. Every Existence or system of the Existence upon which force is not exerted will "accelerate" to the maximal possible speed.

When we say that we apply a force in Classical Physics, in Binary Physics it means that we refrain from exerting a force. That is, by removing the obstruction we eliminate the factor preventing change.

In my extensive work with children, in teaching inventive thinking, I have had many opportunities to talk to them about force. Unfortunately, instances of exerting force on trivial matters in the elementary grades is fairly common.

One of the children once told me that force is when he cannot pass.

He said this, without knowing how right he was.

When I try to move my hand through the wall, I feel that there is a force that prevents me from doing so. And every time when I am about to leave a room, a very powerful force directs my movement in the direction of the door – the walls that blocks me.

I also encounter this force when I am in a traffic jam and the vehicles in front of me slow down my movement…

It seems that observations of the obstructive force are quite prevalent.

We are simply very accustomed to seeing this force in space, and, because we are blind to depth, we are unaccustomed to thinking that the obstruction affects us—and even more significantly—in this dimension as well.

Because the dimensions of space and depth are the same, there is no reason to assume that an obstruction in depth will affect us any less than an obstruction in space.

It also seems that not only does this force seem commonplace, but that it is the fundamental and only force, and all other forces in nature are essentially in fact different interpretations of our perception of this force.

Let us start at the beginning. What is an obstruction?

The Third Rule of Private Binary Physics—the Rule of Non-Merging—states that the Existence can pass only to a cell that contains Placeholder. In other words, when an Existence "tries" to pass to a cell containing another Existence, it is blocked from doing so.

From the moment that the Existence is obstructed from passing to a cell, to which it has been supposed to pass, a change is caused in its motion. It either remains in its place (a total obstruction) or in passes to another cell, since the cell to which it was supposed to originally pass is obstructed (a partial obstruction).

I will give an example. The Seventh Rule—the rule of trivial movement in depth—determines that the Existence moves by default to its most immediate depth. But what happens in a situation where its immediate depth is obstructed? The Existence will be "compelled" to pass to another cell in the space of the next level of depth (this is the Tenth Rule – Gravity).

That is to say that an obstruction causes a change in the movement of the Existence.

Before we go on and discuss obstructions as a fundamental force of private Binary Physics, it is worthwhile to ask ourselves quickly, what is "force" in general?

Classical Physics teaches us that force is anything that can cause a body possessing mass to accelerate.

This definition has no meaning in Binary Physics, since in Binary Physics speed—and also acceleration as a derivative of it—are only illusions that are a product of our mind's interpretation. In the chapter on speed, I will elaborate and explain how this illusion is created in our mind. Currently I will only say briefly by a way of illustration, that since there is only one particle and every cell in the Universe has an identical response time (by the Fifth Rule), then there is only one "speed" in the Universe. All Existence moves at exactly the same "speed".

Therefore, the fundamental force of Binary Physics—the obstruction—does not accelerate the system or the Existence, which it acts upon, but rather opposite – it slows it down. The more significant the obstruction, the more the system of the Existence is slowed down. The most extreme case is the state of the complete obstruction, when the maximum force is applied. In this situation, change/movement of the system of the Existence or an isolated Existence stop completely. In the extreme opposite situation, where there is no obstruction at all (that is, no force is exerted at all), we see the system moving at the maximum possible speed (the speed of the cell's response time), and as we will see later, that this is the speed that is known to us as "the speed of light" in Classical Physics.

In our Private Physics, the rule that creates the obstructing force is the Third Rule – the Rule of Non-Merging. Without it, ostensibly no obstruction would ever be formed.

As a side note, I will emphasize the fact that the obstructing force is a force of our Private Physics, since it is derived from the Third Rule of Private Binary Physics and not from the axioms. It is certainly possible to describe different universes in which the "thing" that causes the movement of the Existence is something else. For example, a double obstruction, or a situation in which the Third Rule of "Non-merging" does not work, and thus "an obstruction" has no meaning. In the second situation, when I move an apple that is obstructed by an apple to the right, the two apples merge into one apple. Does this seem strange? This is only because we have become accustomed to see the opposite. It could be that to another creature living in a universe with no Rule of Non-Merging it will seem very strange that when he pushes an apple, that instead of merging with the apple next to it, it will start to push it away...

We see that because the only physical observation that exists in the Universe is the observation of change (aside from a static observation of the very existence of the Existence and Placeholder), and the only thing that will prevent change (movement of the Existence) is an obstruction, so it is obvious that the obstruction is the fundamental force in our Universe.

The obstruction has the ability to stop or slow down change.

Types of Obstructions

The Second Rule of Private Binary Physics—the Rule of Proximity—determines that the Existence can pass only to one of the surrounding cells in a direct manner (in space and in depth).

In our own Private Binary Physics, every cell is connected to 80 other cells:

- A spatial cube that contains 27 cells in the previous level of depth.

- A spatial cube that contains 27 cells in the next level of depth.

- Its own spatial cube that contains 26 more cells besides that original cell.

The number each cell's connections is of paramount importance. Firstly, only cells that are connected directly to a particular cell will determine its value in the next turn according to the First Rule.

Secondly, the Second Rule allows the Existence in the reference cell to move only to one of the cells that surround it directly.

As a side note, it should be emphasized that General Binary Physics allows for different universes with different arrangements and different numbers of cell connections. Private Binary Physics refers to the three-dimensional space. If you connect to each cell in space, additional cells beyond the 26 that surround it, they would form additional spatial dimensions. It is very difficult for us to imagine how would four spatial dimensions look like, but as part of the axioms of General Binary Physics, by increasing the amount of connections of each cell from 26 to 28 or any other number you can easily generate extra dimensions in space or in depth.

From the quantity and arrangement of the cells that surround every cell in our own Private Physics, certain types of obstructions are derived:

- An obstruction in depth – a situation in which the depth to which the Existence is supposed to move is obstructed.

- An obstruction in space – a situation in which the space to which Existence is supposed to move is obstructed.

- A complete obstruction – all depth cells directly surrounding the Existence—the next direct depth cell and the next spatial depth cells—are obstructed. In such a situation, the Existence is confined and does not move. A state of complete blockage is the only exception to the Seventh Rule, which states that the trivial movement of the Existence is in depth. In this situation the Existence does not proceed in depth and will remain in its current spatial cube.

The Degree of Obstruction

The Existence can be obstructed solely in the depth cells of the successive spatial cube, and can be obstructed in the spatial cubes of a number of successive depth cubes.

As the obstruction gets deeper (i.e. it extends over more spatial blocks), there will be more turns required until it is cleared.

For example, the Existence in a spatial cube d(4)1, which is in a state of total obstruction in the spatial cubes d(4)2–10, will require at least 9 turns until the obstruction is cleared off and the Existence is able to move.

If that Existence is obstructed completely by the spatial cubes d(4)2–100, it will require at least 99 turns until the obstruction is released and it is able to move.

The number of turns that are required for the Existence to be freed up from the obstruction and move is the degree of obstruction.

Of course, the principle of the degree of obstruction is realized also by an obstruction in space in a symmetrical manner.

The Paradox of Obstructions (the obstructive contradiction) and its Solution

When we move an apple in space toward another "static" apple, once the apple reaches the second apple, it is blocked and its movement stops (I am speaking in abstraction of its slow movement towards the second apple, and ignore the discussion of the types of collisions and their consequences). This example describes an obstruction in space that causes the Existence to stop, or according to the rule of collisions, to change its path of movement. This is an observation, which we are familiar with, is comprehensible to us.

Unlike the obstruction in space that is well-known to us, the obstruction in depth is unfamiliar, since we, as human systems, are blind to the depth.

What happens when we have an apple in space that is blocked by an apple in the next depth level?

According to the principles of obstruction that I described above, the apple should remain at its current depth level.

Now suppose that I hold the apple in my hand.

I proceed to the next level of depth (assuming that my system, unlike the system of the apple, is not blocked, of course) under the Seventh Rule, while the apple that I am holding in my hand remains in the spatial cube of the previous depth, because its system is obstructed.

Ostensibly, I expect to observe things that due to the obstruction would remain in depth and would suddenly disappear. As we know, there are no such observations. A paradox has been created.

The solution to this paradox lies in the fact that if the apple in the spatial cube $d(4)0$ is obstructed by an apple in the spatial cube $d(4)1$ and I am not obstructed, when I reach the spatial cube $d(4)1$, I will continue to hold the apple in my hand. But it will not be the system of the apple that was composed of the Existence of $d(4)0$, because that Existence remained there, rather I will hold the system of an apple of $d(4)1$. Understandably, I will not notice this with my blindness to depth, but from a physical standpoint, I will presently hold a system of an apple made from a different Existence.

Of course, if the apple of the spatial cube $d(4)0$ is obstructed by the apple of spatial cube $d(4)1$, then from the very definition of obstruction, the apple of $d(4)1$ should have the same structure as the apple found in $d(4)0$, and be located in exactly the same place in space as is the apple at $d(4)0$.

The above explanation is a fairly abstract clarification of the obstruction paradox, and the full account is more complex. Clearly there are questions pertaining to the moment. What happens if the apple that is obstructing at d(4)1 has a different color? How do I not see that the apple suddenly changes even if we accept the subject of the obstruction? The more complex answer to this question lies in the fact that the system of the apple itself is "smeared" over many spatial cubes in depth, and the pattern of the obstruction in depth is what determines its color…

It should be noted in this context that at very high resolutions, there are particles that emerge and disappear. There is no paradox or contradiction in this. Since our consciousness is blind to depth, we do not interpret this phenomenon as disappearing particles, but, depending on the pattern of particles in depth, we interpret this as different types of materials, i.e. electron, red light, blue light, etc.…

We will see how this happens in the chapter on the formation of matter. The existence of all matter spreads over many turns in depth. The existence of the apple, which I am speaking of, is stretched over many spatial cubes that the human mind interprets as a single spatial cube. Therefore, even when I move in depth I continue to hold the same apple in my hand (because blockage at depth are an integral part of the apple).

The Tenth Rule – Gravity (Bypassing in Depth)

> *"When an Existence encounters an obstruction in the depth dimension, it will circumvent it in a circular movement in a fixed order, to one of the next cells in the depth dimension, while giving precedence to the Existence in the current level in the depth dimension, that his previous in the order. This rule takes precedence over the Eighth Rule."*

The Tenth Rule is most characteristic of the Private Existence Algorithm in our Universe. The Tenth Rule is the source of the birth of orbital motion in space.

When the Existence changes its direction in space due to an obstruction in depth, we human systems, who are blind to the depth, interpret this in terms of Classical Physics as "gravity".

When we look at the Universe at a significantly lower resolution than the resolution of the human system, at the resolution of stars and galaxies, and also when we contemplate the Universe at a significantly higher resolution than the resolution of human systems, at the resolution of electrons and protons, we see how much rotational motion is an integral part of any change in our world. It can be concluded from observations that any change, if we study it at a proper resolution, will start and end in rotation of something around something else.

Application of an algorithm based on the Tenth Rule gives us a universe with such a movement like our Universe.

Why Do We Require a Fixed Order?

This is because the Existence Algorithm must work in such a way so as not to form contradictions, and at the same time it is based on the principle of the cellular automaton, in which all Existence independently receives a "decision" about changing its location; there is no escape from establishing order.

Without a fixed order, a situation can be created when two Existences will pass into the same cell, something that would be contrary to the Second Axiom.

The fixed order of the Tenth Rule is very reminiscent of the "roundabout" (traffic circle) principle. Each car gives precedence to the car that enters the circle from the left… [Each car entering the circle gives precedence to cars that are already moving on the circle.]

Understandably, there is no "left" in the movement of the Existence, but there is a fixed set of priorities. There are 27 cells in depth which can flow into each cell. The first to be given priority is the Existence that is supposed to pass to the next direct level of depth. What about the precise arrangement of the other 26 cells? We need an observation at a higher resolution, perhaps at the fundamental resolution, in order to determine it. Of course it is possible to do a computer simulation for research.

In order to avoid creating a contradiction, we require not only a fixed order, but also an order of precedence. Therefore, these two principles appear in the Tenth Rule.

The Tenth Rule, like all of the rules, is not inevitable in General Binary Physics. It is characteristic only of our own Universe.

If for example we formulate an alternative Tenth Rule, such as "There is no circumventing", we would still receive the change in space, because there would be areas with a more or less significant obstruction, although we would not see circular motion, but only a space that changes in a more static fashion. We would see fairly awkward phenomena like the Sun changing colors without moving from its place.

If we formulate an alternative Tenth Rule, such as "Circumvent always to the right", there would also be a change. We would not receive observations of rotational motion in space. We would see the entire universe moving and spreading in a particular direction.

In short, a change in the Tenth Rule in the context of General Binary Physics creates fascinating examples of alternate universes.

The Tenth Rule Dominates

The Tenth Rule, gravity, dominates over the Eighth Rule—Inertia. We see it in observations as well. When a particle that continues its spatial motion reaches a place possessing gravity (an obstruction in depth), it will change its movement and circumventing in depth takes precedence.

The Uncertainty Principle

According to the Tenth Rule, the Existence circumvents the obstruction in depth in a specific order that is not random at all. In actuality, the way in which the Existence circumvents the obstruction in depth is defined unambiguously.

How, then, is this consistent with the Heisenberg's Uncertainty Principle?

Uncertainty about the position of a particle is due to the fact that the Tenth Rule is based not only on the arrangement of the particles in space, but that the next level of depth, "the future", affects the nature of their movement. Our consciousness is blind to depth, and therefore at the fundamental resolution or in very high resolution, there is a limitation on the level of accuracy if we relate to space alone and ignore the dimension of depth. Of course, this restriction exists only at the highest resolution. At lower resolutions, the statistics of the Tenth Rule balances out and the uncertainty becomes negligible.

Another reason for the uncertainty principle is a technical, not an essential, reason. This reason is also familiar from Classical Physics. The higher we ascend in resolution, the smaller the difference between the resolution of the test apparatus and the test object, and thus the influence of the test apparatus on the test object increases, to the extent that at high resolution the influence is decisive and affects the nature of the test.

Obstruction from the Aspect of Resolution

We see the force of obstruction in any change, not just at the fundamental resolution or higher resolution, but even at low resolutions in general, and at resolutions R(h) in particular.

The most popular example at low resolution for an obstructing force is a traffic jam or waiting in line.

Our instinct suggests that a traffic jam or a queue at the post office are not physical forces. They can be bypassed easily. Well, if it is so easy, from a physics standpoint, to get around them... why does not everyone do it? Why do they occur so frequently? You should think again—more deeply.

The obstruction of the line at the post office shows us an expression of the obstructing force even at the lowest resolutions, such as social systems. Who is not familiar with a feeling that until his boss does not "disappear", he will not make any progress? That is an obstructive force. One Existence obstructs another Existence—that is the power of obstruction at the fundamental resolution—in the fixed universe. Enormous systems of the Existence (a police officer, the route of the road, mountains, the car in front of me) preventing the progress of other enormous systems (my car that is stuck in a traffic jam, and all the cars behind it) are simply a representation of the obstructive force at low resolution.

Inability of utilizing steam power was a barrier to the development of trains. When this obstacle had been removed, the train was developed and with it the limit of speed at which people could get to work was pushed higher, which in its turn was something that eliminated the impediment to settling more remotely from major cities, *etc*. Every "invention" is, from a physical standpoint, a removal of an "obstruction", at low resolution, that enables change.

Because the only physical observation that exists in the Universe is the observation of change (aside from a static observation of the very being of the Existence and Placeholder) and the only thing that prevents change is obstruction, so it is clear that whenever an obstruction is removed, change is made possible.

All the barriers that have been removed during the Industrial Revolution and the Information Revolution (the Digital Revolution) enabled radical changes in the society of human systems, and as a result, significant physical changes to the appearance of the surface (the growth of skyscrapers and big cities; a significant increase of the human population and so on…)

Explosion

Explosion is "rapid release of obstructions in depth on a large scale, that causes a great change to the location of Existence in space".

It is possible to liken the removal of obstructions in depth during an explosion to the situation when we open a dam at once, or suddenly open a barrier that created a huge traffic jam. As a result of this opening of the obstruction in depth, a tremendous change in space is formed, which we interpret as an explosion.

Obstructions and Entropy

In Private Binary Physics, the Tenth Rule appears in space like someone who compresses the Existence, but in essence, in the view of space–depth, the rule causes the Existence to be distributed in space. The circumventing in depth is what causes the Existence to distribute in every direction. At times, the eddies around a large obstruction create smaller obstructions, which we imagine as a decrease in the overall entropy, but over the turns, the Existence is scattered in the Universe, and the obstructions—and the eddies along with them—gradually disappear.

Of course, in General Binary Physics, it is possible to describe universes in which the level of entropy does not change, or in which the level of entropy decreases. For example, a universe with a rule stipulating that an Existence in a cell neighboring another Existence should stick to the latter. In such a universe, it is possible to envision an initial state that will eventually lead to changing all Existence into one huge cluster, and a reduction that leads to the total disappearance of entropy.

Chapter 17
The Nature of Motion

Motion is a mental concept, not a physical one. Change, Existence and Placeholders are physical concepts. Motion is an interpretation.

The only observation which we see, besides the Existence and Placeholder, is change. Motion is a mental concept that reflects the physical concept of change. We do not see an observation of "force". We interpret an observation of a particular motion, or in the language of physics, a certain change, as if it has been caused by a "force". But the observation itself refers to Change.

As a side note, here is justification that the obstruction is a fundamental force. If the only observation (besides the Existence and Placeholder) that we see is motion—change—and there are no observations of any additional type, then it is logical that the only force that will influence the observation will be the force of obstruction—preventing motion—preventing change.

Movement - "a mental concept indicating change of a cell from Existence to Placeholder, and the change of a neighboring cell from Placeholder to Existence."

In fact, motion is a mental interpretation of the Eighth Rule (Inertia), the Ninth Rule (Collision) and the Tenth Rule (Gravity). These three rules alone produce change in space, which a mind interprets as "motion".

The motion of the Existence from one cell to the adjacent cell requires one turn. If the cell to which the Existence is supposed to move is obstructed, movement to that cell can take a number of turns, during which the Existence waits for the obstruction to clear off.

The Rules of Motion in the Fixed Universe—at the Fundamental Resolution

In the following examples, I will use a linear—one-dimensional—space, for the sake of simplicity. Of course, things operate in the same way in two-, three-dimensional spaces, or in spaces with any other number of dimensions. I stress that the number of dimensions in space is determined by the number of connections in space that every cell has in the Binary Field.

The Seventh Rule establishes the most basic motion: trivial motion in depth. Our mind is blind to this motion.

Example 1: The Seventh Rule—Trivial Motion in Depth

Example 1		x	The seventh law		
d(5)	d(4)		0	1	2
0	0	0	1		
		1			
		2			
	1	0			
		1	1		
		2			
		3			
	2	0			
		1			
		2	1		
		3			

Here in this example, the Existence in cell d(5)0d(4)0x1 will pass over to cell d(5)1d(4)1x1 in the next turn, due to the Seventh Rule: Trivial Motion in Depth.

Example 2: The Seventh Rule with complete obstruction in depth

Example 2			Linear Space					
d(5)	d(4)		0	1	2	3	4	
0	0				1			
	1			1	1	1		
	2							
1	0				1			
	1							
	2			1	1	1		
	3							
2	0							
	1				1			
	2							
	3			1	1	1		

Here, the Existence in cell d(5)0d(4)0x2 is in a state of complete obstruction. Therefore, in the next turn, it will not move in depth, and instead will pass to cell d(5)1d(4)0x2. Only in the second turn, after the obstruction has been cleared off, will the Existence pass to cell d(5)2d(4)1x2.

Vector

Vector –

"A physical concept that points the cell which an Existence is supposed to advance in the next turn, in the absence of an obstruction."

The meaning of Vector is that in a situation with no obstruction, the Existence will move not only to the next depth, but also in space, in the direction of the Vector. An Existence with a Vector to the right will move in the next turn one cell forward in depth and one cell to the right in space.

The Vector is an independent piece of data in the cell, along with the Existence.

An Existence without a Vector:

1. Without an obstruction – will move only in depth (as in Example 1 above), or
2. With a total obstruction—will not move (as in Example 2 above)
3. With a partially obstruction - Will receive a Vector in one of two situations:

    (a) In accordance with the Ninth Rule, in a state of collision.
    (b) In accordance with the Tenth Rule, in a state of bypassing in depth (gravity).

The notation of the Vector is in the form of an arrow in a cell that contains the Existence. The arrow is the direction of the Vector.

Example 3—The Eighth Rule—Inertia

The Eighth Rule—Inertia

> *"The default state of a moving Existence in space, is to move in the next turn to the direction to which it moves or was supposed to move in space, in the current turn, while giving precedence in the depth dimension to the Existence before it in the direction of motion."*

Example 3		x	Linear space		
d(5)	d(4)		0	1	2
0	0	—> 1			
	1				
	2				
1	0				
	1			—> 1	
	2				
	3				
2	0				
	1				
	2				—> 1
	3				

In Example 3. we see the Existence in cell d(5)0d(4)0x0 with a Vector pointing to the right. Therefore, since there is no obstruction, Existence will move in the next turn to the cell d(5)1d(4)1x2.

By means of the Vector, continuous motion can be created in a particular direction, and this piece of data permits application of the Eighth and Ninth rules.

To be perfectly honest, I would prefer to put together Binary Physics without the need of the vector. The Vector requires each cell to store two pieces of data: (1) its value – either Existence or Placeholder, plus (2) if the value is the Existence, then it also must contain the vector.

It is simpler and more elegant if each cell contains only one value, either Existence or Placeholder.

In General Binary Physics, it is possible to describe a universe with a different Existence Algorithm, with no need for a vector, since a unique initial state there would allow for complex systems and patterns—identical to those that we see in our Universe—to exist, only on the basis of the location of the obstructions in depth in the first queue. That universe would be based only on the Tenth Rule. At the instant of appearance of such an initial state, there would be no need for the Eighth and Ninth rules. The complexity of such a universe would be that every movement in space (Inertia and Collision) would be required to be "scripted" in initial state. In such a universe, at the frontier of the time (at the highest level of depth with the Existence), there would be no change in space. The change would take place only in the inner layers of depth that encounter an obstruction.

I chose, for now, to add the Vector for two reasons.

Firstly, it allows for description of numerous universes within the framework of General Binary Physics, and there is no need to search for very rare and unique initial patterns in order to explain observations of Private Binary Physics.

Secondly, the explanation of the private Existence Algorithm (the rules of Private Binary Physics) is much easier to grasp through the use of the vector. The rules themselves become more intuitive.

Example 4: Order of precedence of Existence in front of you in the direction of motion

Example 4				Linear Space					
d(5)	d(4)	x	0	1	2	3	4	5	6
0	0		→1	→1	→1				
		1							
		2							
	1	0							
		1	→1	→1		→1			
		2							
		3							
	2	0							
		1							
		2	→1		→1		→1		
		3							
	3	0							
		1							
		2							
		3	→1		→1		→1		

According to the Eighth Rule—Inertia—there is a right of way for the Existence in front of you in the direction of motion (Example 4). Therefore, in the first turn, the Existence marked in green moves forward, and the two Existences marked yellow wait for their turn to move.

As a side remark: The Vector does not pass over in this situation, because according to the Ninth Rule, the Vector does not pass between the cells with identical vectors.

In d(5)3 we see the dispersion of the Existence in such a way that they are found in completely free motion, i.e. in a pure energy state. Such motion is a type characteristic of light rays.

Example 5: The Ninth Rule – Collision

Example 5		x	Linear Space				
d(5)	d(4)		0	1	2	3	4
0	0				→1	←1	
	1						
	2						
1	0						
	1			←1			→1

The Ninth Rule: Collision

> *"When an Existence is found in a cell next to another Existence, if they have different default motions, the direction of movement will pass in the next turn, from one to the other, in the direction of movement."*

By the Ninth Rule, two Existences that are found in adjacent cells exchange their vectors with each other in the direction of motion.

We learn that in the framework of one turn, two processes are performed: the exchange of a vector, and after that, the movement itself (there is a third process: checking that there is no conflict, that we will see immediately below).

In the fixed universe (in the fundamental resolution), collisions are only elastic. Therefore, as soon as there is a collision, the vector passes from one Existence to another, thus continuing its movement.

If there are only elastic collisions in a fixed universe, what is a plastic collision?

A plastic collision is a mental concept. It is an illusion, which results from an interpretation that an observer gives to a collision at low resolution and at a very slow speed. In such circumstances, systems change their shape (deform/join) and heat is released (the Existence is released due to the impact and moves in directions that seem, at low resolution, to be random).

What the observer would see if he looks at the collision at R(0) and at the maximal speed of time, would be an elastic collision. The Existence in the area that belongs to one system is compressed into a cloud of Existence in the area that belongs to a second system, and part of the Existence flies to the sides of the region of the two bodies (which is heat).

Example 6: The Ninth Rule – The non-transfer of the vector not in the direction of motion

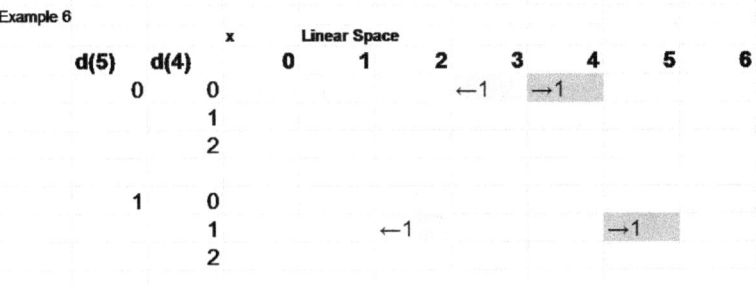

Since both Existences in d(5)0 are in opposite directions, there will be no exchange of the vector between them, in accordance with the Ninth Rule.

Example 7: The Ninth Rule – Destructive Interference

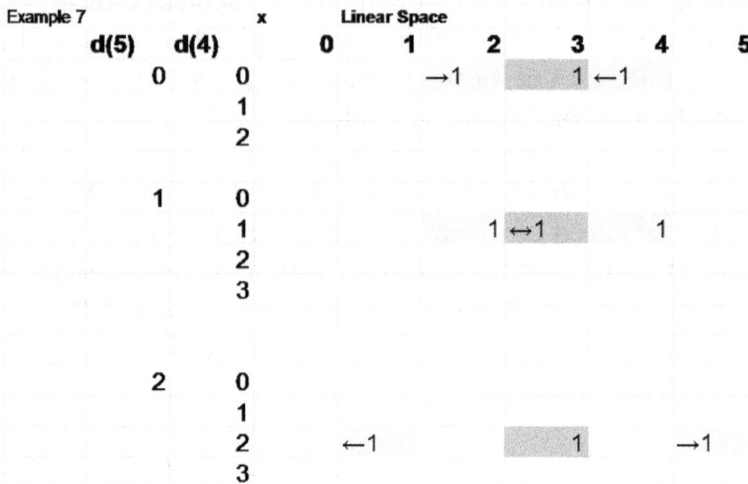

In cell d(5)1d(4)1x3 we see a double vector, which reflects destructive interference.

Destructive interference and its continuity are evidence that a cell can accommodate a two-way vector.

Example 8: Checking to Avoid a Contradiction

In the example above, we see that within the turn, another action is required—checking for contradictions.

The two Existences at d(5)0 tend to reach cell d(5)1d(4)1x3, according to their vector.

Between the two Existences at d(5)0, there is no transfer of information, according to the First Rule.

This creates a contradiction.

One possible solution to this matter is that the cell d(5)1d (4)1x3 performs a check to avoid a contradiction (and indeed within the framework of every turn in each cell). If the cell "sees" that Existences from two different cells are supposed to reach it, it will cause the Existences in the original cells to reverse their vector.

It is likely that there is a more elegant solution than the mechanical calculation of checking for a contradiction. This point should be improved as part of the research.

Thus, the following three processes are performed within the framework of each turn:

1. Transfer of the vector, according to the Ninth Rule.
2. Test for a contradiction.
3. Change.

Example 9: Newton's Cradle

Example 9		x	Linear Space					
d(5)	d(4)		0	1	2	3	4	5
	0	0		→1	1	1		
		1						
		2						
	1	0						
		1			1→1		1	
		2						
		3						
2		0						
		1						
		2			1	1	→1	
		3						

Based on the Eighth Rule—Inertia: Note that every Existence gives the right of way to the Existence in front of it in the direction of motion.

Newton's Cradle is the name of the apparatus that can be found in any mans gift store. It has a number of metal balls, each tied separately in a row, to a wooden rack. When you lift up a ball on one side and release it to strike the other balls, the ball on the opposite end goes up, and then back again.

By the Ninth Rule of Private Binary Physics—the Rule of Collision—the first ball passes its vector to the next ball which obstructs it. The next ball passes it to the next ball, and so on, until the vector reaches the last ball, which is not obstructed, and it lifts up in the direction of the vector, according to the Eighth Rule—the Rule of Inertia—until it encounters an obstruction in space or in depth (which is interpreted as gravity).

Why does not the last ball lift up before the vector reaches it?

At first glance, we would expect that the last ball (let's say, the rightmost one) would lift up regardless of the vector—because it is not obstructed on the left.

The last ball is definitely moving regardless of the vector. It moves in depth in accordance with the Seventh Rule. As a result of the Sixth Rule, motion in depth does not require movement in space (this is as distinct from movement in space, which requires movement in depth). Therefore, the last ball moves in depth, but does not move in space. (It is understood that I am speaking of the ball as a system at low resolution. At higher resolutions, such as in the resolution of the atom, the Existence that comprises the ball moves in space **and** in depth.) The moment that the vector reaches the last ball, it receives, in accordance with the Ninth Rule, also spatial movement in the direction of the vector, and persists with it, in accordance with the Eighth Rule—and therefore it goes up.

We should differentiate between a vector, a potential change, and change. A potential change and change exist inherently according to the Seventh and Eighth rules for every Existence. The vector is not a desire to change, or change itself. The vector is the direction of change..

A physical observation that illustrates the Eighth Rule well is the light rays that continue in their direction as long as no force is exerted on them, i.e., they are not obstructed.

Example 10: The Seventh Rule – Trivial Movement in Depth.

	x	0	1	2
d(5)	d(4)			
0	0		1	
	1	1		1
	2			
1	0			
	1		1	
	2	1		1

	x	0	1		0	1
d(5)	d(4)					
0	0	1				1
	1		1		1	
	2					
1	0					
	1	1				1
	2		1		1	

Example 11: The Tenth Rule – Gravity

	x	0	1	2
d(5)	d(4)			
0	0		1	
	1		1	1
	2			
1	0			
	1	←	1	
	2		1	1

d(5)	d(4)	x	0	1	2	0	1	2
0	0			1			1	
	1		1	1			1	
	2							
1	0							
	1				→ 1		← 1	
	2		1	1				1

In the above examples, we see how the vector (motion) is formed, due to the influence of the rule of gravity.

> *Gravity – "A mental concept which notes the change in direction of motion of an Existence, due to an obstruction in depth."*

We interpret the obstructions in depth as gravity, because the mind is blind to depth.

The Tenth Rule states that the bypassing will be in a circular motion in a fixed order. Therefore, in a linear space, the motion will be performed in the following order:

- If the next level of depth is vacant, then to that level.
- If the next level of depth is obstructed, then to the depth on the left side.
- If the next immediate level of depth and the level of depth on the left side are obstructed, then to the depth on the right side.

The determination of right and left is my own arbitrary determination—the exact prescribed order should be investigated.

Example 12: The Effect of the Tenth Rule on Existence with a Vector

d(5)		d(4)			0		1	2
	0		0			←	1	
			1		1			1
	1		0					
			1				1	
			2		1			1
			3					

d(5)		d(4)		0	1	2		0	1	2
	0		0		←	1			←	1
			1				1		1	
	1		0							
			1	←	1				1	
			2			1			1	
			3							

Tenth Rule – Gravity (Bypassing in Depth)

> *"When an Existence encounters an obstruction in the depth dimension, it will circumvent it in a circular movement in a fixed order, to one of the next cells in the depth dimension, while giving precedence to the Existence in the current level in the depth dimension, that his previous in the order. This rule takes precedence over the Eighth Rule."*

You can see in the above examples the states when, due to obstruction, the Tenth Rule—Gravity—causes the vector to disappear and/or to change its direction.

Since the Tenth Rule takes precedence over the Eighth Rule, it is what forms the vector (when it causes Existence to perform a bypassing in depth, while changing its spatial location) or negates it (when, due to an obstruction in depth, the Existence is not able to move in space).

It is important to note that the disappearance of the vector and/or the change in its direction are caused only in the direction of the obstruction, that is, if there is a double vector "above" and on the "right", and the obstruction is on the right, the vector above will remain and continue to influence the Existence, as can be seen in Example 14 below.

Example 13: A Double Vector in Two-Dimensional Space

		y					
d(5)0	d(4)0		5				
			4				
			3		← ↑ 1		
			2				
			1				
			0				
		x		0	1	2	3

		y					
d(5)0	d(4)1		5				
			4				
			3				
			2				
			1				
			0				
		x		0	1	2	3

		y					
d(5)1	d(4)0		5				
			4				
			3				
			2				
			1				
			0				
		x		0	1	2	3

		y					
d(5)1	d(4)1		5				
			4	← ↑ 1			
			3				
			2				
			1				
			0				
		x		0	1	2	3

We see in this example how the double vector affects the motion of the Existence upwards and to the left.

Example 14: A Double Vector in Two-Dimensional Space in a State of Obstruction

d(5)0	d(4)0	y					
			5				
			4				
			3		← ↑ 1		
			2				
			1				
			0				
		x		0	1	2	3

d(5)0	d(4)1	y					
			5				
			4	1			
			3				
			2				
			1				
			0				
		x		0	1	2	3

d(5)1	d(4)0	y					
			5				
			4				
			3				
			2				
			1				
			0				
		x		0	1	2	3

d(5)1	d(4)1	y					
			5				
			4		↑ 1		
			3				
			2				
			1				
			0				
		x		0	1	2	3

d(5)1	d(4)2	y					
			5				
			4	1			
			3				
			2				
			1				
			0				
		x		0	1	2	3

In cell d(5)1d(4)1x1y4, we see the Existence that came from cell d(5)0d(4)0x1y3.

The vector "left" of the Existence in cell d(5)0d(4)0x1y3 disappeared due to the obstruction because of the dominance of the Tenth Rule. The vector "above" remains.

We see an example of motion of this type at low resolution in observations when we fire a cannonball, and it is affected by gravity (the Tenth Rule) and moves down, while its vector is preserved in the direction at which the cannon has fired.

Motion according to the Tenth Rule will not change the direction of a vector, which is not in the direction of the obstruction. In this way, circular motion can continue at the fundamental resolution, and in parallel, motion at low resolution of the entire system in a particular direction. Since the vector which is not in the direction of the obstruction is preserved, a circular bypassing can be performed in depth, and at the same time the Existence maintains its "will" to move in the direction of the vector, which is not in the direction of the obstruction.

As a side remark, and to conclude the topic of motion, I will say that it is possible to define the entire motion of the Existence in a fixed universe—at the fundamental resolution—by means of a network of logic gates between the cells.

The Ninth Rule versus the Tenth Rule—A Contradiction and Its Resolution

The Tenth Rule (Gravity) is an exception to the Eighth Rule (Inertia) (and as a derivative of this, to the Ninth Rule (Collision). The Tenth Rule dominates it. When an Existence encounters an obstruction in depth, it acts according to the Tenth Rule, while negating the default motion of the Eighth Rule in the direction of the obstruction. According to the Tenth Rule, the Existence bypasses the obstruction in depth in a circular manner.

We clearly see in observations how the Tenth Rule takes precedence over the Ninth and Eighth rules in the difference between the straight motion of a light beam (free motion that is free of any obstruction in depth) and the circular motion that occurs within matter, i.e. mass (motion that is based on obstruction in depth).

The Rules of Motion in Space

The Eighth (Rule of Inertia) and Ninth (Rule of Collision) rules are what determines the basic motion in space. These are the "default" rules.

Only the fundamental force—the obstruction—can change the default motion in space either by a full stop (complete obstruction) or deceleration (partial obstruction) or by action of the Ninth (Collision) or Tenth (Gravity) rules.

Motion due to an obstruction in depth is when an Existence being obstructed in depth tries to bypass the obstruction. It performs this in a motion that is reminiscent of the movement of the knight in chess (in depth and sideways):

	1	
2	Existence	4
	3	

The above diagram shows which cells the Existence will proceed in the next depth, if the immediately following depth is obstructed. This is just an illustrative example—the exact arrangement should be determined in further research.

Since the Existence Algorithm must operate continuously and without contradictions, when each Existence autonomously calculates its course in the next turn, there is no choice but to establish a uniform direction of bypassing for each Existence.

This is similar to vehicles entering a traffic circle. Each one knows that it gives the right of way to the next car to the left, and the direction of travel is to the right. Thus all of them travel coherently. Every car is required to perform a minimal calculation: to check either there is a car coming from the left, then it is obstructed; or there is no vehicle coming from the left, then it is able to move forward.

There is no choice but to establish a similar rule for bypassing in a situation of obstruction; otherwise we will have a situation in which two Existences will appear in one cell, and we will have a contradiction to the Second Axiom which determines that every cell can contain only one of the elementary particles.

If we were computer programmers who want to build a universe that flows without contradictions, we would be required to make this rule.

In fact, the Tenth Rule is the archetype of the most famous type of motion in the universe—circular motion.

How are obstructions in depth and the cellular system consistent with our observations of continuous motion?

The low resolution at which our mind sees the observations is what allows it to give the interpretation of continuous motion.

Our cognitive psychological time includes a vast quantity of turns. Therefore, a relatively significant obstruction, in space or in depth, is required in order that we can feel it. A significant obstruction in space will be interpreted by us as a wall, for example, and a significant obstruction in depth as gravitation.

Why does motion on the quantum level appear random in Classical Physics?

This illusion is created as a result of the nature of the Tenth Rule activity and the mind's blindness to depth.

As the mind contemplates the Universe at a higher resolution, it will see "random motion" of particles.

Albert Einstein stated that "God does not play dice", and, as usual, was right.

Random motion is only an illusion.

As resolution grows towards the fundamental resolution, more the mind finds it difficult to determine the exact location of the Existence in the next turn.

Because the exact location of the Existence is determined by the Tenth Rule, according to obstructions that appear in the next depth, and our consciousness is blind to the next depth, thus due to the lack of information our mind interprets the new location of the Existence as randomly positioned.

But when we read the Tenth Rule, we understand that this is not random motion at all. The Tenth Rule explicitly states that motion is performed according to a fixed order!

When we decrease resolution, the problem is solved on its own. It is like observing a seemingly random motion of a drop in a wave as opposed to the motion of the entire wave, which appear regular to our mind. At low resolution, consciousness sees a wide section of depth. Although it interprets it by compression as one substance and not as depth, but it still sees a wide section of depth. As the resolution decreases and we observe a wider section of depth, the smaller is the illusion of randomness.

Chapter 18
Systems

System and System Algorithm

System—"A mental concept. A pattern of two or more Existence"

Except for the physical concepts, Binary Physics relates to everything as a system—the system of the electron, quark system, proton system, atom system, molecule system, heart system, human system, automobile system, television system, computer system, society system, country system, Planet Earth system, the solar system, galaxy system and the largest possible system—the system of the universe in general.

System Algorithm—"A mental concept. The regularity of motion of the system at low resolution. Based on a superposition of the regularity of all of the movement of Existence that comprises the system at the Fundamental Resolution."

Note: The concept of "motion" relates to space and to depth. Even in a situation of no motion (total obstruction), it is still part of the regularity of motion.

From a physical standpoint, everything in the Universe possessing regularity of motion that contains more than one Existence, is a system.

Once there is regularity at the fundamental resolution, then perforce at any higher resolution, regularity will dominate, constituting a superposition of the regularity at the fundamental resolution. Therefore, from the instant that the Existence Algorithm dominates at the fundamental resolution, then at any observation of a system at any resolution, a system algorithm derived from the Existence Algorithm will dominate.

To be clear – a system cannot exist with a regularity that is not based on the original regularity of the Existence Algorithm. It is understood that when observing the system at a very low resolution, we see a very complex regularity, which is a superposition of a huge quantity of Existence. This regularity can seem to us entirely different than the Existence Algorithm. But this is merely a false impression.

Each system is characterized by three fundamental features:

1. The number of the Existence and Placeholder particles that comprise it.

2. The way the Existence and Placeholder are distributed in space–depth – the pattern of the system.

3. Its system algorithm = its regularity.

The more that the system is composed of more than one Existence and, on the other hand, the more that we study it at a lower resolution—the more the complexity of its system algorithm increases. That is, there is a direct correlation between the amount of the Existence that comprises the system and the complexity of its system algorithm, and an inverse relation between the resolution of the system and the complexity of its system algorithm.

The system algorithm allows the observer to see the entire system at low-resolution as if it is a single unit, which has its own rules of motion.

The Existence Algorithm is a physical concept. The System Algorithm is a mental concept, because it does not stand alone, but is a unified mental interpretation of many runs of the Existence Algorithm, so that the mind imagines that it is dealing with an independent algorithm (regularity), which operates on the system.

To facilitate the understanding of the concept of the system algorithm, let us compare it to the system itself. The system itself does not exist. The system is a mental concept within which the mind interprets in its imagination a large amount of Existence at a low resolution, as if it is a single unified thing, the system. Just as we watch TV and see lots of pixels as if they were a picture of one thing.

Exactly the same principle operates at the algorithmic level. A system algorithm does not exist. A system algorithm is a mental concept in which the mind interprets a lot of interactions of the Existence Algorithm at low resolution, as if it were a new algorithm that stands alone and influences the whole system in a consolidated fashion.

Just as a combination of a huge quantity of the Existence will form, at low resolution, a system of a flower, which is an infinitely more complex system than an isolated Existence, similarly a combination (superposition) of a tremendous amount of interactions of the Existence Algorithm will create, at low resolution, the illusion/interpretation of very complex algorithm operating on an entire system.

The lower the resolution, the more the mind considers a larger quantity of interactions of the Existence Algorithm in a consolidated fashion, and so will the System Algorithm seem in observations to be more complex.

I would like to emphasize – there is no algorithm that has been specifically designed for the benefit of the operation of a particular system.

An environment that enables multiple interactions that stems from the Existence Algorithms (an environment that does not have total obstruction ; has sufficient Existence to create interactions between them ; and there is an Existence Algorithm that allows "joining" of Existence particles: one Existence tracking after another, or rotation of one Existence around another Existence, *etc.*) will lead to a situation, in which very complex patterns of motion of systems will develop at low resolution. These patterns will be mentally interpreted at low resolution as a specific regularity that operates on systems.

Human systems are an example of a highly complex system algorithm, which at very low resolution appears as one type of regularity that operates on the system, while, at the fundamental resolution, constitutes a huge quantity of interactions of the Existence Algorithm.

How is this possible? Just as you can create simple algorithms that will, over the course of their operation, create spectacular graphical patterns for which the algorithms themselves were not planned originally, similarly will the Existence Algorithm, after a sufficient number of turns and in a particular environment of distribution of Existence, form complex patterns, such as human systems.

Another example of how this is possible: I am able to create an amazingly simple algorithm that will create random text pages by changing the arrangement of letters and words in a given book. Over enough turns, this algorithm will produce the Bible. And the New Testament. And the Quran. In a split second, the creation of a masterpiece will appear from this algorithm. Similarly with the Universe, which creates countless patterns as a result of the Existence Algorithm, occasionally masterpieces—such as human systems— appear. On second thought, there is a reason to reconsider whether human systems deserve the title of "masterpiece"...

If the specific algorithmic environment does not just allow for the formation of complex system algorithms such as human systems, but even gives them a survival advantage, then these system algorithms will begin to be more common.

The system algorithm of human systems exists at a very low resolution. Therefore it is infinitely more complex than the system algorithm that is found at the basis of the atom.

Most systems that we see in observations, large as well as small—atomic systems, planetary systems or star systems—possess a relatively simple system algorithm. This is a system algorithm in which it is easy to identify within it the reflection of the Existence Algorithm, and in particular the influence of the Tenth Rule—Gravity (Circumventing in Depth).

In observations, there are only a few systems that have developed very complex patterns of the system algorithm – such as human systems.

The system algorithm of an isolated Existence is the Existence Algorithm.

We descend in resolution, and it is possible to see the system algorithm of the atom. This system algorithm is the manner in which the system of the electron spins in depth around the system of the proton, in accordance with the Tenth Rule. This system algorithm is already based on a huge number of interactions, each being individually based on the Existence Algorithm.

Descending in resolution, we can imagine the system algorithm of the molecule.

Descending in resolution, we can imagine the system algorithm of an element.

Descending in resolution, we can imagine the system algorithm of a compound.

Descending in resolution, we can imagine the system algorithm of the heart of a human system.

Descending in resolution, we can imagine the system algorithm of a human system, which is of course a very complex system algorithm based on an enormous and incomprehensible amount of interactions of the Existence Algorithm.

This is similar to the difference between looking at high resolution at one logic gate, and the very limited output that it can have, and looking at low resolution at a huge quantity of logic gates and treating them as a single unit, which we imagine as a new i7 Intel processor able to perform according to a very complex logic for every input that is entered into it.

About Resolution, Chaos and Order

Since the Existence Algorithm controls the movement of all Existence without exception, by definition there cannot be a chaotic system. In any configuration of Existence that we observe at any resolution, there must be some regularity that obtains.

If so, why do we see "chaotic" systems in observations?

1. Our minds have developed evolutionarily so as to identify regularity only in systems that assist our survival. Therefore, our mind succeeds in easily identifying certain patterns, and has great difficulty, or even does not succeed in the task with other patterns. For example, our mind will easily identify regularity in the motion of the system of a deer that it wants to hunt, and will have difficulty identifying the regularity of a system of carbon atoms in the atmosphere.

2. Our mind observes the universe at a particular resolution R(h). It is easy for it to identify if there is regularity at this resolution. There are systems for which defining their regularity is very complicated at our resolution, but is simple to define at another resolution. Since we are confined to our resolution—we have difficulty in defining the regularity in these systems, and mistakenly call them "chaotic".

3. There are systems whose regularity (their System Algorithm) is complex to such an extent that the calculational ability of our mind is not strong enough to define it.

Order and Chaos are not separate forces. The private Existence Algorithm that causes the scattering of Existence from the point of the Big Bang (Chaos) is exactly the one that cause - at the same time as the scattering, and due to the obstructions that were formed - the formation of fascinating stable patterns of Existence (Order), such as human systems.

Stable Systems

> *A Stable System— "A system possessing a pattern that does not undergo self-destruction. A system where the internal order of Existence within it matches the character of the Existence Algorithm such that its pattern is preserved throughout two or more turns (Existence does not "flee" from it, and the ratio of Existence to Placeholder is preserved)."*

It is understood that stable systems that we see in observations, are relatively stable. The more the pattern of obstructions and spinning of Existence within the system allows it to survive over more turns, the more will the system be considered stable. The lifetime of the system is what reflects the strength of its stability. For example, the lifetime of human systems is approximately 80 years. Nickel, for example, has one isotope (59Ni) whose half-life is more than

10,000 years, as opposed to three other isotopes (56Ni, 57Ni and 66Ni) whose half-lives are less than ten days. It is understood that the definition in years or in seconds is a relative, not an objective, definition, and is influenced by the speed of the system. If we want to speak of the objective length of the lifetime of the system, we must speak in terms of turns.

There is no difference between the stability of a system of a soap bubble and the stability of human system, other than the number of turns that they survive.

How is a stable system formed?

If we think about the definition of a stable system, we see that the condition for it to form is that its pattern should match the Existence Algorithm. The more that the character of the system's pattern will match the Existence Algorithm—the more the system is expected to survive and to be stable.

The definition of stability is also dependent on the resolution. That is, the character of the system that we study at the Fundamental Resolution must match the Existence Algorithm. The character of a system that we contemplate at low resolution—must also match the system algorithm of the system that contains it. It is understood that the definition of stability of a system as compared to the system algorithm, and not the Existence Algorithm, is only a matter of convenience and simplicity, because in the end every system algorithm is derived from the Existence Algorithm.

For example: If we consider the system of a fish, located in the system of the ocean, at low resolution, we will see it as stable, surviving. The pattern of the system of the fish matches the System Algorithm of the system of the ocean. The system of a fish thrown into a desert system will not survive, since it does not match the System Algorithm of the system that surrounds it.

Another example, this time at high resolution: in a particular universe, the Existence Algorithm determines that every Existence will proceed one cell to the right in each turn, unless there is an "Existence" that obstructs it. In the initial state, we are given the following pattern of Existence (a pattern of the type Existence/ Placeholder/ Existence/Placeholder/Existence...):

0101010101000000000000000

The character of this pattern fits the given Existence Algorithm, and it can advance to the right through the length of the turns without collapsing. In the next turn, the pattern will appear as follows (every Existence moves one cell to the right ; the pattern of the Existence is preserved):

0010101010100000000000000

Let us describe another initial state in which we are given a pattern of "Existence" as follows (a pattern of the type: 2 Existences/ Placeholder/2 Existences/Placeholder/2 Existences):

0110110110000000000000000

The character of this pattern does not fit the Existence Algorithm and is expected to collapse (in terms of failing to maintain its pattern) already in the next turn.

In the next turn, the pattern will appear thusly:

0101101101000000000000000

This is no longer the same pattern. We saw that, in order that a stable system will be formed, it must fit the Existence Algorithm of the private universe in which it is developing.

There is a direct correlation between the stability of systems and their prevalence. When a particular system randomly develops that matches a particular environment—it survives. Therefore, we will start to see in that environment more and more stable systems of that kind that survived.

Stability of the system is a derivative of its suitability to the Existence Algorithm of the universe in which it operates. Therefore, in the framework of General Binary Physics, it is possible to describe systems possessing a particular pattern that survived in a universe in which one Existence Algorithm is operating, and will decay under an Existence Algorithm that operates in another universe.

The beginning of evolution (Development of Stable Systems in Private Binary Physics)

In order that stable systems will develop in our Private Binary Physics, their system algorithm must be such so as to be able to preserve its pattern during movement of Existence that arises from the Ten Rules of the private Existence Algorithm.

From the initial state onward, very many systems have developed in our universe. Most of them have survived for only several turns, because their pattern was not suited for our private Existence Algorithm.

From the moment that the macro-process of scattering of Existence started in accordance with the Existence Algorithm, the micro-process of survival started. That is the birth of evolution. Therefore, the beginning of evolution was not in the development of life, but rather already at the first turn—at the initial state. When the first turn occurred, and with it the change—the process of survival of systems started. As a derivative of this, an evolutionary process of requiring increasingly complex system algorithm, in order that the pattern of the system would remain stable – started.

The first stable systems that survived were systems of electromagnetic waves.

These are systems possessing a relatively simple pattern, lacking mass and thus lacking internal obstructions in depth (we will see in the chapter about the birth of materials and the nature of mass, that internal obstructions in the system are mass).

Due to the fact that systems of electromagnetic waves lack any internal obstructions, their system algorithm is simple, and does not have to deal with the Tenth Rule—Gravity—within their system (They are influenced by the Tenth Rule as an obstruction external to the system).

Because the patterns of electromagnetic waves are simple and it is easy for them to survive, a very great variety of systems of electromagnetic waves was formed (microwaves, ultraviolet light, visible light, infrared…). Every different frequency of the electromagnetic wave constitutes a different pattern in depth of Existence and Placeholder. In the chapter, dealing with the birth of the materials, I will present the depth patterns of electromagnetic waves.

Compared to the patterns of electro-magnetic waves, the patterns of particles - for example, the electron and proton - contain internal obstructions in depth within their system (and therefore they possess mass). From the instant that a particular system contains an internal obstruction in depth, it is forced to deal with the internal spinning that is formed within the system due to the Tenth Rule. The Tenth Rule states that every Existence that is obstructed in depth by another Existence, tries to circumvent it in a circular motion. This circumvention causes the Existence to change its direction, and according to the Eighth Rule (Inertia), to start scattering in every direction, and thus dismantle its pattern…

Therefore, in order for such a system to survive, it must possess a very unique pattern. This pattern must form a system algorithm, that will move the Existence within it in such a way, that every time that Existence from within the system will be in an exit path from the system, another Existence from the system will obstruct it (according to the Ninth Rule—Collision), and will bring it back into the system.

Because, such a unique pattern is required in a state in which there are internal obstructions in the depth of the system —particles with mass (possessing internal obstructions) are extremely rare in our universe.

Indeed, alongside millions of patterns of systems of electromagnetic waves (each frequency band of electromagnetic wave is a separate system with a unique pattern in depth of Existence/Placeholder), we see only three common base systems that possess a pattern that contains mass (internal obstruction): the proton, the neutron and the electron.

Physics today speaks of a jungle of particles. Despite this, it is definitely possible to speak of the three above systems, as systems that constitute the foundation of matter. Most of the other particles which are discussed are secondary particles of these particles.

A Complex System - "A Stable System Possessing internal Obstructions in depth (Mass)"

Particles with mass are complex systems. This is because their mass is our mind's interpretation of the internal obstructions in depth within their system.

The level of complexity of the system depends on the pattern of arrangement of obstructions within it, and not the quantity of obstructions. It is possible to have a system with an enormous quantity of obstructions that are arranged in a simple periodic structure, and thus its level of complexity will be low (its system algorithm will be relatively simple). An example of such a system is the iron nucleus of a star. On the other hand, it is possible to have a system with a relatively small quantity of obstructions, that are arranged in a unsystematic manner, and thus its level of complexity will be high (its system algorithm will be relatively complex). An example of such a system is a human system.

In the chapter on the Birth of Materials, I will present examples of patterns of systems with mass - electrons, protons... And then things will be clearer

Parenthetically, I will note that in General Binary Physics there is a huge number of universes with no stable systems developed in them at all, either due to their initial state, or to their Existence Algorithm, or both. The question arises: Out of all the possible universes, how precisely in our Universe did stable systems develop, that advanced to the level of complexity of conscious human systems? This is a question based on circular reasoning. After all, only in universes in which the Existence Algorithm and initial state allowed for the development of complex systems will there be those who will ask this question...

In the first turns in the queue, when the density of Existence due to the particular initial state of our existence was at its peak—practically no stable systems developed. This is because the obstructions were so significant that the Existence almost didn't move.

In our queue, when on one hand there are still many obstructions, but very few total obstructions—fascinating stable patterns of motion of Existence develop. A good example of complex patterns that formed in our queue are those of human systems.

The more the dispersal of the Existence became extensive, the more the systems had to "find" more and more effective techniques to survive. As the techniques become more efficient, more system algorithms become not only more stable, but also more complex. In our Private Universe, the dispersal continues to spread, and gradually only systems with a particularly sophisticated system algorithm will succeed in surviving.

In terms of General Binary Physics, in a universe in which the Existence can move back and forth in depth, there may well be a situation when powerful system algorithms will develop that attract more and more Existence towards them, and thus stop the dispersal, and perhaps even change the trend from dispersal to gathering… (As a side note, it is possible in General Binary Physics to describe a universe with the initial state of maximal dispersion, whereas it Existence Algorithm is an algorithm of gathering…)

Evolution and survival of systems are mental, not physical, concepts. There really isn't development of systems algorithm or survival of systems. There is only the movement of the Existence in accordance with the Existence Algorithm. Our mind imagines patterns of the Existence as systems, and picture the Existence Algorithm at a low resolution as regularity, which operates the systems. Therefore the development of systems and the survival of one system and the extinction of another are mental interpretations of the regularity of systems at low resolution.

The system algorithm does not "want" to maintain its system. But simply just as soon as it does not maintain it, the system algorithm will cease to exist.

At low resolution, system algorithms start to develop. Only certain system algorithms survive – those that manage to maintain their system over successive turns. The mind interprets these algorithms retrospectively as "having a will" to survive. A will to become organized. The truth is that there has been no such a "will", but rather a lack of choice. Algorithms without such "will" have not survived, and therefore only systems with a "will to survive", growing more and more in its sophistication, remained. History is written by the victors…

The example of clouds is what I would like to use to illustrate this.

Sometimes we look at the sky and see clouds of different shapes. We see a cloud in the form of a horse chasing a rabbit-shaped cloud. Our mind does not only imagine systems of clouds, but even imagine a regularity in their movements.

Of course, a horse-shaped cloud is not trying to chase a rabbit-shaped cloud.

The horse-shaped cloud also does not try to survive.

Its pattern is formed and survives over a certain time period. As long as there is a pattern, our mind interprets it as a regularity that acts on the entire system of the cloud.

A similar example is system algorithms which—we imagine—control the stars that make up the signs of the zodiac. Each system of a "zodiacal sign" in the heavens is an imaginary system, which survives over time and obeys a particular regularity at low resolution.

The only principle that directs the system algorithm is its survival.

The system algorithm itself has no desire to survive or not survive. As can be seen from the Existence Algorithm (the progenitor of the System Algorithm), in its ten rules, there is no mention of a rule of survival.

Ostensibly we have encountered a contradiction. How do we see only systems with a system algorithm fighting for survival of them, while at the same time we establish that there is no rule that requires the system to survive ?

The answer is incredibly simple.

We see only system algorithms that succeeded by chance, and continue to succeed, to survive in an environment in which the

Existence is scattered in every direction.

Our mind gives an interpretation only to system algorithms that survived. Supposedly the system algorithms that have survived are the ones that we see in the observations. System algorithms that have not survived do not appear in observations. All Chaos is also a kind of pattern—where the difference lies between it and the pattern that we call system—is that in the chaos pattern our mind doesn't able to indentify any regularity (a system algorithm).

Our mind makes the following inference: If there is a system, and the system exhibits regularity that is called a system algorithm, then the system survives thanks to its system algorithm. This is an incorrect inference.

This is because we derive the assumption that the system exhibits regularity from the belief that if there is a system, then there is a specific regularity that operates it. Just as there is no specific regularity in a horse-shaped cloud, there is no specific regularity in any other system.

All those system algorithms that have not managed to survive – we do not see. Probably over the course of the turns a vast quantity of different and varied patterns of Existence flows developed, that at a low resolution appear to us as different and varied system algorithms.

From this huge number, only few patterns have survived, and these are mainly those that comprise atoms…

A very small fraction of the patterns continues to evolve into more and more complex examples. As more turns transpires and the Existence is being scattered more and more, the more difficult it becomes for patterns to survive.

As I mentioned, several of those very complex models that have survived and continue to survive are the patterns of human systems and the system algorithm that operates them.

Because human systems and the system algorithm that conducts them are no different, from a physical aspect, from any other system or any other system algorithm, the only motivation of human systems is survival. Not because they wish it, but rather because they will cease to exist at the instant when the system algorithm of human systems stops succeeding in allowing them to survive.

The common strand among physics – chemistry – biology – history is the development of more and more complex systems, with increasing survival ability in an environment where the Existence is being scattered in every direction.

Physics – the system of the Atom.

Chemistry – the system of the molecule, matter.

Biology – a system with an active reflex for survival.

History – social systems.

All four fields represent how, as we decrease in resolution, the systems increase in the amount of Existence that they contain—thus by necessity a more complicated System Algorithm is formed within them.

Durable Systems

Another central matter with which stable systems must deal is the Force = An external obstruction to the system.

When a system encounters an external obstruction, generally its depth pattern will collapse. In order that this will not happen, the system is required to possess a unique pattern, which contains an internal spin in such a way that it will prevent its collapse in a case of an external obstruction. For example, the systems of the proton and the electron must be stable in such a way that when they encounter a gravity field (an external obstruction in depth) they will not collapse.

> *Durable System* — *"A stable system that her pattern is built in such a way, that the internal spinning of Existence within it, prevents external Existence from entering into it or causing its decay"*

The system algorithm of a durable system does not only move the Existence in it in such a way that prevent the Existence within the system from "fleeing" from it, but also in a way in which the Existence spins within the system prevents external Existence from entering within. In a durable system, there is no "exposed" Placeholder. At all times, another Existence obstructs its regions of Placeholder and ensures that external Existence will not enter within it.

Most systems in the universe are not stable, and certainly not durable by themselves. The existence of most systems depends on their interaction with other systems.

In our private Binary Physics, only very specific arrangements of patterns of obstructions in space-depth will allow the formation of stable or durable systems.

As a side point, if a purely durable system will be formed (that is absolutely durable), there will be no way, in the framework of the axioms, to dismantle it, and from the moment that it is formed, it will continue to exist forever.

At times, systems succeed in achieving stability or durability only by means of "cooperation" with another system. The spin of the two systems around one another protects both from penetration of an external Existence amongst them—something which protects their pattern from collapse.

From the moment that a system is formed at a particular resolution, whose connection to another system will cause them to survive in a better manner, an exponential process will begin by joining together small systems (with respect to the number of Existences that they contain) into larger and more complex systems.

Particular atoms survived better by merit of their joining with other atoms. This led to the development of a stable system of greater complexity—the molecule.

Particular molecules survived better due to their joining with other molecules, and thus a stable system of even greater complexity developed-matter.

An Evolutionary Process = Certain patterns that are formed randomly are more suited to the Existence Algorithm (and at low resolution, certain patterns are more suited to the system algorithm that encompasses them) = From patterns that possess suitability to the Existence Algorithm, stable and durable systems develop = Stable and durable systems by definition survive longer than other systems = An increase in the frequency of the stable and durable systems, and the development of these stable and durable systems.

Why does the atomic system not break down at the time of a collision between two systems at low resolution (as for example two balls)?

The atomic system is based on the spinning of the Existence around an obstruction in depth, in accordance with the Tenth Rule. The system of the atom is based on the unique pattern of arrangement of the Existence in space and in depth, in such a way that even at the time of collision it does not come apart. A description of this pattern is called "the Binary Atom". The atom is relatively a stable and durable system.

The pattern of the system of the atom does not only form a system algorithm that moves the Existence within it in such a way that it prevents Existence from within the system from "fleeing" from it, but also the way in which the Existence spins within it prevents external Existence from entering within. In a durable system, there is no "exposed" Placeholder. At all times there is another Existence that obstructs its regions of Placeholder and ensures that external Existence will not enter within. Thus the system is preserved in a state of collision. The pattern of the system of an atom is preserved over the course of the collision, and only the vector in the direction of the collision passes on from one system to another.

Of course, out of all of the systems that can be described, we require a system possessing a pattern of the Existence that is very unique, in order that spontaneous decay of the pattern will not occur, nor perish in a situation of collision with another system.

And certainly we can say that we are speaking of a rare system. The atomic pattern is unique, and besides the atom, we are not familiar with another pattern that allows for the existence of stable systems with internal obstruction (mass), which we call matter.

In General Binary Physics, it is possible to describe universes in which there is an Existence Algorithm that allows forming stable systems on an extremely large scale. For example an Existence Algorithm that determines that every Existence starts to follow another Existence from the moment that this Existence is located to its right, is an example of an Existence Algorithm that allows for creating very large stable systems.

It is possible to say that also in our private universe, the Existence Algorithm allows for the formation of huge stable systems that are based on an enormous quantity of Existence—human systems, through countries, up to galaxies…

Continuity of the System

Since the system algorithm of every system is derived from the Existence Algorithm, and the First Rule of the Existence Algorithm (the Rule of Autonomy) states that the value of every cell is determined on the basis of the values of the cells surrounding it, the systems must be continuous.

Of course, the matter of continuity is not obligatory, and in General Binary Physics it is possible to describe non-continuous systems. That is, a system algorithm that states, for example, that if a cube of Existence to the right side of a universe moves to the right, then a cube of Existence in the left side of this universe will also move to the right, in such a way that there will be a regularity that connects the movement of the two systems. In such a situation, it is possible to see one non-continuous system in the two cubes.

If such a situation would prevail in our Private Universe, then when we would move half an apple in Israel to the right, suddenly we would see half an apple in China moves to the right…

In observations, we see that systems are based on the contiguously of the Existence to one another. Molecules are not formed from atoms from the opposite end of the galaxy, but rather from atoms nearby.

Therefore, it is possible to state that in Private Binary Physics, a condition for forming a system is the relative proximity of the Existence.

I say "relative proximity", because the intention is proximity on a level that allows one system to influence a second system, and the First Rule requires that in the basis of the influence should be the proximity of the two systems.

Proximity is dependent on resolution, since the entire concept of the system is a mental concept. As mentioned above, there actually isn't any system, but rather the mind imagines a collection of Existence as if it is one system. Therefore, at higher resolution, the mind relates to Existences that are relatively close to each other as one system, and at a lower resolution, the mind can relate to Existence that is spread out over a vast expanse as one system.

The condition of proximity does not determine that we must speak of adjacent cells in the fixed universe—at the fundamental resolution. The Tenth Rule at low resolution will cause us to see large flows of the Existence spinning around other large flows of the Existence, where eventually our consciousness observing all this at a very low resolution will interpret the flows as a stable, uniform system possessing one form of regularity.

A side-note: It must be remembered that proximity is required not only in space, but also in depth.

The system of the Earth is an example of a relatively simple system algorithm of bypassing in depth. Human systems are more elaborate examples of bypassing in depth.

However, the system of the Earth is vastly larger than a human system (from the aspect of the quantity of Existence that the system contains and the scope of its distribution), but there is no connection between the size of the system and its complexity—that is, the complexity of the arrangement of its obstructions. The more complex a system, the more a sophisticated system algorithm is required, in order for it to survive.

Development of a successful system algorithm for bypassing in depth is what ensures the preservation of the system and its survival. In our private binary physics a system, which "does not find" a system algorithm to help it perform successful bypassing in depth, will decay. It is possible to compare the Earth to a car moving in a traffic jam, and the advanced system algorithm of human systems to an agile scooter that moves ingeniously between the cars and bypasses the traffic jam in a most sophisticated manner.

In the definition of Time Resolution, it was determined that the mind contracts different cyclical patterns in depth, and interprets them as different materials in space. At low resolution, this is true regarding systems as well. Due to the cyclical nature that characterizes systems, there is no way, for example, to say where exactly in depth the human system "Josef" begins and where it ends… There is an overlap between the different areas of the system, and in fact it is possible to say that the system "Josef" is stretched out in depth and repeats itself…

Resolution of Systems

As I mentioned earlier, as opposed to the Existence and the cell which are physical concepts, the system is a mental concept. There is actually no system, and there are actually no rules by which it acts. There is only a great amount of the Existence, and there is only the Existence Algorithm.

A mind that observes the Universe at a low resolution imagines a pattern composed of a great deal of the Existence as a single system, and the superposition of all the behaviors of the Existence according to the Existence Algorithm as a single complex system algorithm.

Resolution of Systems:

The lowest resolution system
The system of the universe.

Very low resolution systems
The system of Black holes.
The system of a galaxy.
The solar system

Low resolution systems
The system of the Earth
The system of a country
The system of human society
The system of history
The system of economics
The system of a city
The system of a bank
The system of a Company

Resolution of Human Systems R(h)
Human Systems.
A car.
An apple.

High Resolution systems
The system of a molecule.
The system of an atom.
The system of a proton, Quark and electron.

Physical concepts in the fundamental resolution R (0)
Existence.
Placeholder.
Cell.

From a physical standpoint, there is no difference between the imagined system of a car and the imagined system of a society, bank or a country. All concepts are mental, i.e. they exist only as interpretations in the observer's mind. Both concepts operate according to a certain regularity (a system algorithm), which is a complex superposition of a huge quantity of the Existence that acts according to the Existence Algorithm. There is no car that travels somewhere, exactly as there is no State of Israel. Both are our low-resolution mental interpretations of the Universe. What is this car, from a physical standpoint? When we increase in resolution we get a steering wheel, doors, an engine and windows; we increase further in resolution and we get plastic, glass, iron; we increase even more in resolution, and we receive atoms of the elements; we increase even more in resolution, and we receive quarks and electrons. We increase even more in resolution, and we receive pattern of Existence in depth that obeys the Existence Algorithm. All the components of the car together at low resolution obey the regularity that constitutes a superposition of all behaviors of the Existence that composes it.

Even the principles of the systems are maintained regardless of the resolution. High resolution systems, like the systems of atoms, are examined in two parameters - their stability (the pattern itself is not collapsing) and durability (pattern that can prevent external Existence from penetrating and cause collapse). Low resolution systems, as states, are being evaluated in exactly the same parameters. This is because all systems are the same essence and the resolution is only an illusion of consciousness.

Why our minds start to imagine systems?

Once the private Existence Algorithm and the private initial state have been determined, an autonomous process of the flow of Existence in the universe (the First Rule) starts. According to the private Existence Algorithm and private initial state, many patterns of the flow of the Existence in space–depth started to develop—different and varied patterns, of which the vast majority did not survive the turns of the queue until our days. Among the mixture of patterns, some started to appear with greater frequency. These patterns were characterized by the system algorithm that managed to preserve them in a reality of dispersal of the Existence in that turn and in that area (stable systems). Eventually, after a huge number of turns, we arrived at the current turn in the queue in which our consciousness observes the Universe, and at this advanced turn there is a very limited (relatively) quantity of stable systems, which are familiar to us from the Periodic Table of the Elements. It is reasonable to assume that many turns ago, when there were many obstructions and it was very easy for stable systems to be formed, the Periodic Table of the Elements contained additional elements. It is also reasonable to assume that in many more future turns, as the Universe develops even more as a result of the action of the Existence Algorithm, we will see less and less stable systems, and the Periodic Table will be curtailed.

Within this autonomous process, our mind, which studies the process at a low resolution, imagines that certain patterns of the flow of the Existence (that repeat themselves) are systems possessing an independent existence. Our mind imagines an apple as a system with an independent existence in the same way in which it imagines a cloud in the form of a rabbit as possessing an independent regularity.

Our consciousness works out of a large system (our body) that can be defined only at low resolution. For this reason, our consciousness has become accustomed to observe the Universe at a low resolution, identical to that of the system that contains it.

We see the system of an apple simply because the resolution of this system is suited to the survival of the system that contains our consciousness. Of course, at high resolution, apples are composed of a tremendous quantity of the Existence. But for the needs of humanity's survival, our consciousness is not required to understand the internal system of the systems at high resolutions, but rather it is enough for them to recognize their regularity at low resolution, that dictates that the system called "an apple", when it is plucked from the tree, is picked as a single unit.

As a side point, I will note that, by means of analyzing "big data", it is possible to artificially identify, system algorithms at low resolution, which we did not develop evolutionarily in order to identify. There is no difference between a system algorithm of a "horse" and a strange system algorithm that is revealed in "Big Data", except for the fact that we do not naturally perceive the latter.

Since the existence algorithm controls all, in each cluster of Existences necessarily rule regularity. Therefore, other creatures, who evolutionary developed to see the reality in different resolution - will see a completely different systems when they will look at the same reality.

The System Algorithm and resolution

As we have seen earlier, the rules of systems are mental concepts, as are systems themselves.

The rules that govern every system are derived from the superposition of all the interactions of the Existence that compose it.

As the level of resolution Decreases, mostly, the level of complexity of the system algorithm gets higher. The reason for this is that every algorithm, at low resolution, is composed of a huge number of interactions of the Existence Algorithm in the fundamental resolution. As we descend in resolution, the algorithm is getting composed of more interactions of the Existence Algorithm, and therefore it becomes more complex.

It is possible that there are specific cases in which, despite the system being at low resolution, due to the specific pattern of Existence within it, it is possible to summarize its system algorithm trivially. But these are rare and exceptional instances.

The regularity of the wave system movement in the sea is infinitely more complex than the rules of motion of every molecule of which it is composed.

The Example of a Drop of Ink

To explain better the nature of systems and the nature of the system algorithm that operates as their basis, I like to give the example of a drop of ink.

When we drop ink into a glass of water, the ink drop begins to develop into spectacular forms. Within these shapes, our consciousness imagines different systems, like rings, ellipses, snakes...

Over a number of turns (which increases if the liquid in the glass is more viscous and contains many more obstructions), these systems will endure. The ellipse that I observe while I am writing this has already lasted for, according to my subjective time, at least 5 seconds, and in objective terms, an incomprehensible number of turns.

As my mind imagines that there is a system—an ellipse, similarly it imagines that the ellipse has a regularity of its own.

This imagination is not a far-fetched one. At low resolution, there really is a system of an ellipse, and this ellipse really has a regularity that operates it.

Of course, when we look at high resolution, then we will see that the system essentially consists of many molecules of the Materials of the ink that collide with many molecules of water, and the system algorithm is essentially the basic rules of physics that are familiar to us, that operate at the level of molecules.

At low resolution, our consciousness can certainly recognize the regularity of the ellipse. Consciousness can determine, with a high level of certainty, that the ellipse moving to the right will continue moving to the right the next second as well.

What Essentially is Physics?

Physics:

"A science that investigates the Existence Algorithm, and comes to provide the answer to the question: "What will be the observation of a pattern of Existence in another X turns, given the initial state of the present turn?"

The First Rule of Private Binary Physics shows us that this question has an Ambiguous answer at the Fundamental Resolution.

If this is physics, then certainly it is possible to say that from a physical standpoint, we can relate to the regularity at low resolution—such as that of the ellipse—irrespective of the fact that this regularity is derived from another regularity at higher resolutions.

The regularity that we identify in the ellipse is, from a physical standpoint, the system algorithm that operates as its foundation.

An additional (and final) example, in order to understand the nature of the system algorithm as an independent regularity in its own right at low resolution, even though it is derived from another regularity at a higher resolution, can be seen in the ocean waves. Once we see a particular wave rise in the sea, over a certain period of time, we are able to analyze its regularity at the low resolution of the entire wave, and realize that it is going to strike us, even though we can not calculate the location of each and every molecule in the wave.

It is important to note: A system algorithm is an algorithm that changes throughout the lifetime of the system, and at the end of the existence of the system, it fades away and disappears. This is distinct from the Existence Algorithm, which is continual and constant.

Rules of Motion of systems

Of course, our minds do not see motion at the fundamental resolution, but rather at the low resolution R(h).

Because of the low resolution at which our mind observes the Universe, we see motion not of an individual Existence, but rather of huge systems of it.

> *The motion of a system – "a series of changes caused by the Existences possessing a vector in a uniform direction over a number of turns, which the mind interprets at low resolution by means of superposition of all of the changes as one big change in a particular direction."*

When we see a car moving to the right, what we really see is a huge group of cells that their value varying from left to right from the Placeholder to Existence, while maintaining a fixed pattern, which we imagine to have a certain form that we call "a car".

Go and learn: Binary Physics consists of change of the cells values. There is no "motion". When large groups of cells change their value while maintaining their pattern, our mind interprets this as motion. The central reason that allows us to see the motion of systems is the low resolution, at which we observe the Universe.

The human mind, which is blind to the depth, relates to the concept of motion only as motion in space. A system that moves only in depth is interpreted by our mind as a static system, despite the fact that from a physical standpoint, there is no doubt that this is a system in motion. From the physical aspect, the definition of motion is identical to motion both in space and in depth. Due to the Seventh Rule—Trivial Motion in Depth—in our Private Binary Physics, a truly static system (that does not move) is only a system that is found in a state of total obstruction (in depth). In General Binary Physics, in a state when the Existence Algorithm does not contain a rule such as the Seventh Rule, it would be possible to find also systems that are not in complete obstruction, and still are static in the physical sense of the word.

Physical Rules at Low Resolution

Just as systems are huge collections of the Existence that seem to be one pattern at low resolution, so it is with the rules.

The ten fundamental rules—that together make up the Existence Algorithm of Private Binary Physics—seem totally different when they are studied at lower resolutions.

Because the systems are entities at a low resolution, also the rules that direct them are at low resolution.

At low resolution, a tremendous quantity of interactions that result from the ten basic rules appear as one complex rule. The rule at low resolution reflects a superposition of a vast number of fundamental interactions.

There is an inverse relationship between the resolution and the complexity of the rule. The lower the resolution, the more complex will be the rule of motion.

A physical rule at low resolution – "*a mental concept that is an interpretation of an observation at low resolution on a large quantity of changes that result from the Existence Algorithm and performing a superposition of the changes into a single rule.*"

The system algorithm is a physical rule at low resolution.

Chapter 19
Change (Energy)

There are several definitions of energy in Classical Physics, but it appears to me that this one is the most popular of them all: "A scalar physical quantity that indicates the quantity of work that can be performed by a force".

I prefer another definition from Classical Physics that simply seems more clear to me: "the ability to convert work into heat." In converting energy into heat, it is possible to utilize the energy in its entirety. Heat increases the extent of movement of particles in space. The hotter it is, the more the particles move, and more energy there is.

Binary Physics defines energy simply yet comprehensively: "change".

The intention of "change" is the scope of change in location of the Existence in the Universe at the current turn as compared to the previous turn. That is, how the Universe will appear from $d(6)x$ to $d(6)(x+1)$.

What is change?

> Change – "The number of exchanges of the Existence with Placeholder, in a given space–depth".

A distinction should be made between **change** and **difference**.

> *Difference - "The disparity between the distribution of the Existence and Placeholder in the Universe between the initial state (or the reference turn) and the final state ($\times$ turns after the initial state or the reference turn)"*

It may be that there is no difference, but there is much change. That is, there can be a lot of energy that will perform many changes, but eventually the initial state and the final state will be identical. For example, a coiled spring possessing potential energy gets released and goes up and down until it stops. The difference (a released spring) does not reflect all of the changes that the spring has performed until it has reached an entirely relaxed state.

What is the source of change?

The source of change, and hence the source of energy, is in the Existence Algorithm.

The rules of the Existence Algorithm, especially the Seventh, Eighth and Tenth rules, are those that bestow the Existence with its energy; that is, the will to change its position and thus to create change.

Were it not for the Existence Algorithm, no change would be created in the Universe from the turn of the Initial State. A universe without an Existence Algorithm is a static universe without energy.

Because the source of change is in the Existence Algorithm, all Existence inherently contains one unit of change.

This unit of change can be actualized when the Existence moves during a turn, or remains within the Existence as potential for change when the Existence is obstructed from moving.

How does Binary Physics define potential energy?

Potential energy – "potential for change"

In a state of lack of obstruction, the change is maximal. In a state of total obstruction, the potential for change is maximal.

Potential for change can occur in one of two cases:

1. An internal obstruction in the depth of the system (mass). When that obstruction opens, the change will be actualized. The greater the quantity of the obstruct Existence in the system (the mass), the greater is the potential for change. This situation is called in Classical Physics "the energy that is latent in matter". In fact, when we detonate a nuclear bomb, we are releasing an obstruction in depth of plutonium, creating a lot of movement of the Existence that causes a tremendous change, which we call energy.

2. An external obstruction of the system (gravitation/height). The situation that we call "the height" is a condition, when the obstruction has no expression in space, but appears in depth. A high region is a zone with an external obstruction to the system in depth, and a low region is a zone, where there is a smaller external obstruction in depth, or there is no external obstruction at all. The picture that emerges in space in these situations is confusing to our consciousness that is blind to the depth. The reason for this is because an image of the appearance in space is a mirror image of the obstruction in depth. Where there is no external obstruction in depth, all the Existence of the system will attempt to pass, and then an obstruction will be formed in space, and where there is an external obstruction in depth, the Existence will not accumulate in space, since due to the Tenth Rule, the Existences will try to circumvent the obstruction, and thus flow to a region with no obstruction. It can be said that the external obstruction that is concealed in depth creates a type of pattern that the Existence in space is poured into it, and

creates a "negative" image of it. In any event, in a high location/gravity location, the Existence encounters an obstruction in depth, and therefore it "wants" to move to a lower place with no obstruction in depth. The potential for change is formed when the "Existence" that is obstructed locally in a region with a significant obstruction in depth (a high place). If the Existence is obstructed locally in a high place, the moment that the local obstruction opens, it will flow in accordance with the large obstruction of that region—which changed it into a region that we call "high"—and its potential for change will turn into change. When the system "descends", its internal spin in depth is in the general direction of spin in its area (as if it flows with the current in a river). When the system "ascends", its internal spin in depth is opposite to the general direction of the spin in its area (as if it flows against the direction of the current in a river). I will expand and explain the whole thing in the chapter "The Secret of Gravity".

The total energy in the Universe (both the actual and the potential) = Total quantity of the Existence. Each Existence contains a unit of change (if it is moving) or a unit of potential for change (if it is obstructed).

In the Initial State, just before the Big Bang, there was the maximum potential for change in the Universe. There was the maximum obstruction in depth and in space.

As more turns transpire, obstructions in the Universe, both in space and in depth, are freed, and as a derivative of this, the potential for change became change. In the language of Classical Physics: the potential energy of the Universe decreased and the free energy in the Universe increased.

Theoretically, if the Universe continues to expand, eventually all the Existence will be dispersed in every direction in space and in depth, all the obstructions will be opened up and the Universe will be only change (pure energy), and there will be no more potential for change (mass = potential energy). Stable and durable systems, which due to their pattern, do not self-decay, are the exception to this.

The change is influenced by two factors:

Obstruction – how much is the Existence free to move in the next turn. The more there are obstructed Existence in a particular turn the potential change will be greater and the change will be smaller. For example, a complex system of matter possessing an internal obstructions (mass) will possess a higher potential change and a low amount of change, when compared to a light ray that is lacking internal obstructions (massless), that possesses a high amount of change, but has no potential change.

The ratio between the Existence and Placeholder. Since we treat Existence as a particle, then we can say that the more Existence there is, the greater is the change and the potential change.

Now I will explain how each factor affects the change.

How does obstruction affect change?

An energetic state is a condition with a large amount of change. The most energetic state is a situation with no obstruction at all, and a lot of Existence advances in depth and in space at the maximum possible rate of change, that is derived from the response time of the cell. Our minds interpret the maximum possible rate of change as movement at the maximum possible speed (which we call the speed of light) and as pure energy.

The rule is that the more a system has internal obstructions in depth (that is, more mass), the slower its movement, and thus it generates less change (i.e., it is less energetic). And *vice versa*: The less a system has obstructions in depth, the faster its movement, and therefore it generates more change (i.e., it is more energetic).

A system without internal obstructions at all generates the greatest amount of change, and will be pure energy (without mass). An example of a system without internal obstructions in depth is an electromagnetic wave.

We will see in the chapter on the birth of materials that a light wave is actually a movement of the Existence with obstructions neither in depth nor in space. Since the internal obstruction in depth is what is interpreted by us as mass, and is also what causes the rate of change to slow down (like a traffic jam in which the cars move slower than the maximal speed at which they can technically travel), we understand why our mind interprets a light wave also as massless, also as a system that proceeds at the maximum possible speed, and also as possessing the greatest amount of change (the most energetic state in the language of Classical Physics).

A less energetic state is a condition in which, due to obstructions, change is lower and the potential change is higher.

The more the obstruction in depth dominates, the more mass is formed and less energy is there (or, more correctly, when there is more mass, there is the same amount of energy, just in the form of potential energy that is stored in matter). Conversely, when the obstruction in depth is smaller, the mass decreases and is "converted into energy". We see here the identity of mass and energy as dependent on the pattern of the Existence in depth. In other words, we receive here a confirmation of the observation in Classical Physics of the mass–energy identity. When the Existence flows freely in depth, it is interpreted as pure energy without mass—an electromagnetic wave. When the Existence proceeds slower due to obstructions in depth, it is interpreted as a state with mass (a particle), and as a state with less energy (or, more accurately, as a state possessing potential energy, or chemical energy stored in matter). Of course, when we exert a force and we open the obstruction—for example, by a collision between Existences that causes them to scatter in all directions and open the blockage—the obstruction is released and the Existence starts flowing in depth at an increasing speed, and the matter converts to energy.

A system with full obstruction does not create any change at all, and it is pure mass (lacking energy and full of potential energy, which I will discuss shortly). An example of a system that is fully obstructed in depth is a system that is at the temperature of absolute zero.

A state completely without energy is a condition with zero change—where the Existence has nowhere to move, there will be no change over turns in the queue.

Here is an example:

The most energetic state possible—free energy—is an electromagnetic wave:

Pattern of a point in depth R(0) – 10101010101

The particle above moves at the maximum speed. It is encounters no obstruction, and therefore it is massless. It possesses the maximum energy.

In this pattern, all the Existence can advance one cell to the right (for purposes of illustration, let us assume that the direction of advancing in depth is to the right), since the cell to the right of every Existence contains Placeholder.

A state of the greatest potential energy that is stored in the mass—is the most dense matter possible (with the highest specific gravity):

Pattern of a point in depth R(0) – 11111111111

The particle above is absolutely frozen (at absolute zero), and shows no movement at all. It is devoid of speed. It possesses the maximum mass.

In this pattern, the Existence cannot move, because the cell to its right contains another Existence— and it is obstructed.

According to the mass and the energy of every particle that we see at R(h), we can reconstruct its system pattern in depth at R(0). The greater its mass, the more the clusters of obstructions of the Existence in depth will seem longer. The higher its energy, the greater the regularity of spaces will appear between one obstruction and another.

We saw that the potential change increases, and change decreases, when there is an obstruction in depth. In the language of Classical Physics, the greater the obstruction in depth, the more potential energy there is, and less free energy.

Another example:

High energy = High level of change = a pattern in depth that appears as follows: 000100010001

The Existence can flow freely and at its maximal speed, that is, at the response speed of the cell.

Our mind will interpret this condition as massless (because mass is an internal obstruction in the depth of the system), and as a situation with movement at the maximal possible speed = the speed of response time of the cell = the speed of the electromagnetic wave (the speed of light).

In the next turn, all of the Existence in the example will continue to flow in depth in accordance with the Seventh Rule (for the example we will assume that the Existence "wants" to moves in depth from left to right):

The state will be 0000100010001

Low energy = low ability to change = obstruction = potential energy (potential change) = pattern in depth that looks like this:

01110011100111

The Existence cannot move freely, since it is obstructed by another Existence, and it cannot violate the Third Rule, the Rule of Non-Merging, which forbids it from passing to a cell that contains another Existence.

Our mind will interpret this state as a condition in which mass exists (because mass is an internal obstruction in the depth of the system), and as a state with movement at a speed slower than the maximal possible speed; despite that all Existence can theoretically pass to the next cell at the response speed of the cell, it delays and waits before the cell becomes vacant—something that creates change at a slower speed in the system (like a traffic jam).

In the next turn, the Existence continues to flow according to the Seventh and Third rules, and the situation will be as follows (for the example we will assume that the Existence "wants" to moves in depth from left to right):

011010110101101

We see clearly that the Existence—as a pattern in depth—flows much more slowly here.

In fact, the obstruction in depth is a single essence, that our mind interprets in three aspects:

First, the obstruction in depth and in space, when it's external to the system, it's interpreted as the fundamental force that stops the change, or alternately, the opening of the obstruction (a lack of obstruction) that intensifies the change.

Second, the obstruction in depth, when it's internal to the system, it's interpreted as the illusion of mass. When a large quantity of the Existence is "stuck" together and cannot flow freely, our mind interprets this as mass.

Third, the obstruction is interpreted as potential energy. When the obstruction is high or total, the potential change is high and change is low. When the obstruction is low or nil, the potential change is low and the change is high or maximal.

Note: If the entire Universe were in a condition of total obstruction with no movement—and thus no change—there would be no reason to speak of energy. In a situation with no obstruction in the entire Universe, change would be influenced only by the quantity of the Existence.

Summary:

A pattern possessing a high level of obstruction = potential for change = potential energy (from a high location/gravitation) or chemical energy (from the pattern of the matter) = a low level of change from one turn to the next = a small change (if at all) in the location of the Existence in every turn in space and in depth = the brain interprets (at low resolution) the pattern with "long" obstructions of the Existence in depth as mass = a (relatively) cold state = for example, an ice cube.

A pattern possessing a partial obstruction = partial potential to actualize change = a high level (energy) of change from one turn to the next = a significant change in the location of the Existence in every turn in space and in depth = the brain interprets (at low resolution) the pattern with "short" obstructions of the Existence in depth as a mass that moves quickly = a hot state = for example, steam.

A pattern without obstruction = total actualization of potential for change = change (energy) = maximal level of change from one turn to the next = a significant change in the location of the Existence in every turn in space and in depth = the brain interprets (at low resolution) the pattern as being without obstructions of the Existence as massless = pure energy = electromagnetic waves.

How does the Existence/Placeholder ratio affect change and the potential for change?

We see that also in states lacking obstruction, different levels of change (energy) can be formed.

We see this clearly in observations when we compare electromagnetic waves (that are without obstructions, without mass) and have different frequencies.

As we will see in the chapter on the birth of materials, frequency of the Existence in the depth pattern is what differentiates between a less energetic red light beam and a more energetic blue light beam.

Suppose that we ask why blue light has much more energy than red light.

Since the properties of substances, including their energy levels, are determined in depth, we have to examine what is the difference in the pattern of the Existence in depth between blue and red light electromagnetic waves.

In order to characterize the pattern of the system (matter or energy) in depth, it is required to answer two sub-questions:

Does its pattern possess mass? = What is the extent of the internal obstructions that it contains?

What is the total energy of the system? = potential change + the change = since for every Existence, there is one unit of change or potential for change, what is the quantity of the Existence in the system? = What is the relation between the Existence and Placeholder?

In the states of blue and red light, there are no obstructions. We know that there is no obstruction in the light, since both move at the maximal possible speed, which is the derivative of the cell's response time. When there is an obstruction in depth, the speed of motion of the wave will be lower than the maximal speed, which is the derivative of the cell's response time. We also know from observations that light waves are massless, and therefore without obstruction.

What about the ratio of the Existence to Placeholder?

We know from observations that the frequency of blue light is greater than the frequency of red light.

The Sixth Rule of Private Binary Physics, the rule of the space–depth symmetry, requires that the frequency in space be identical with the frequency in depth. Reminder: The frequency of the Existence in depth is the ratio of the Existence to Placeholder in the depth pattern.

For example

A pattern of a point in depth of red light at R(0) will appear as:

000010000100001

And a pattern of a point in depth of blue light will appear at R(0) as: 001001001001001

Of course, the ratio of the Existence and Placeholder in these examples is for illustrative purposes only.

We see that the frequency of the Existence in depth of blue light is higher than the frequency of the Existence of red light.

As the frequency of the Existence in depth increases— more Existence comes into play—the ratio of the Existence to Placeholder increases in favor of the Existence and change intensifies (energy intensifies).

Ten children can make more change in a specified amount of time (a turn) than one child, and a hundred children can create more mess (= change) than ten.

As there is more Existence, thus the change that can be created from one turn to the next increases.

Difference

Difference, as opposed to change, is the disparity between the initial state and the final state, independent of the number of turns that occurred between them. The difference is influenced by a number of factors:

The Existence to Placeholder ratio: On the one hand, more Existence allows creating more difference, and thus the Universe appears to possess development of more complex systems, while on the other hand, too much Existence will create a state of excess obstructions in the Universe, which hampers the formation of difference.

The number of turns that occur between the initial state and the final state—the more turns, the greater the possibility of forming greater difference.

The number of cells to which each cell is connected. If in our

Private Physics, we speak of a spatial cube possessing 27 cells + 27 more cells in the next level of depth + 27 more cells in the previous level of depth, then the potential for a fundamental difference of each Existence is 27. This number is derived from the Second Rule (the Rule of Proximity), the Sixth Rule (Time – Symmetry of Space–Depth), and the Seventh Rule (The Rule of Trivial Motion in Depth), which define the number of cells to which the Existence can flow in a state of the absence of obstructions. In General Binary Physics, in a state of a great quantity of the Existence without obstructions, the potential for difference will be the number of Existences to the power of the number of connections that each Existence has, subject to the limitations of movement of the Existence Algorithm of the Universe.

The Existence Algorithm: an Existence Algorithm different from that of our Universe, which does not include, for example, the Fourth Rule (Conservation) or the Second Rule (The Rule of Proximity), will allow for much greater differences in a lower number of turns.

About Heat and Cold

Let us start with the Binary Physics definition of heat:

Heat: "Change".

As we can see, the definition of heat is completely identical to the definition of energy. The reason for this is that we are speaking of two identical concepts. Both heat and energy are a movement of the Existence, that is, change. When there is no movement of the Existence—due to either absence or/and obstruction of the Existence—there is no change, and therefore there is no energy and no heat.

Heat is a mental concept, not a physical concept. Change is a physical concept. Heat is not a property of matter, but rather our mind's description of the movement of the Existence.

More heat = more change.

Less heat (cold) = less change.

In fact, heat is actualized energy = change. Cold is potential change (obstruction) or lack of energy (lack of the Existence).

Is the Existence itself warm? If it is in motion, the answer is yes. If it is obstructed, the answer is no.

Heat is our name for the Existence in motion – free Existence.

Therefore, we can say:

Free Existence is hot Existence.

Obstructed Existence is cold Existence.

From here we arrive at the following definitions:

Maximum cold = absolute zero = "complete blockage in a specific area, or a specific area that contains only the Placeholder" = Total lack of change.

Maximum heat = maximum movement = "Maximum Existence in a specific area without obstruction" = Maximum change = Maximum frequency of the Existence in depth, and, as a derivative of the Sixth Rule, also in space.

We see that Binary Physics determines the absolute lower and upper values of cold and heat.

And from these definitions, the scale of cold–heat is also derived.

More Existence and less obstruction in a defined region = more movement of the Existence = more change = more heat

The lower the resolution, at which our mind studies the observations, the wider the scale will be, because there will be many more intermediate states between complete obstruction and a maximum of freely moving Existence.

Is an electromagnetic wave hot?

If we have defined heat as change = movement of Existence, then the electromagnetic wave contains heat.

As long as the system of an electromagnetic wave floats freely among Placeholders and does not collide with other Existence, our mind will not sense heat.

The moment that a system of the electromagnetic wave will collides with a system possessing internal obstruction (mass), then due to the Ninth Rule (Collision), an Existence that collides with another Existence within the obstruction will cause them to move in the next turn in the direction in which they collided, and thus they will agitate the obstruction and essentially open it, and automatically, as a derivative of the Sixth and Seventh rules, additional Existence will start to move and a movement of the Existence will start = change will start to form = heat will be generated.

An example: A system of An ice cube that contains an internal obstruction in depth. The Existence of the Sun's rays hits the Existence of the obstruction. Then this Existences will start to move according to the Ninth Rule, and the obstruction will start to open. In a chain reaction, the Existence that has been obstructed starts to move according to the Sixth and Seventh rules, and a process of change will be created = a process of movement of the Existence in the region, which will intensify itself until the complete dissolution of the obstruction in depth. The observation that our consciousness will see in space: a ray of sunlight that hits an ice cube and melts it.

Another example: A traffic jam due to an accident. Each car is an Existence. The traffic jam itself is an obstruction. All the cars are stationary = no change = cold. A policeman arrives (= a ray of light). He moves the car that suffered the accident (the Existence of the policeman moves the Existence of the car, according to the Ninth Rule). The obstruction is freed up.

Begins growing movement of cars = change increases = heat increases. This continues until the complete dissolution of the obstruction: the traffic jam clears off.

Our mind prominently sees the "heat" during the break-up of an obstruction (mass), because then a relatively small area undergoes a lot of significant change = a lot of movement of the Existence.

To summarize:

Heat = Change = motion of the Existence = absence of obstruction/reducing the level of obstruction = energy.

Change from the aspect of General Binary Physics

In terms of General Binary Physics, it is possible to imagine universes in which the scope of change (the energy spectrum) will be greater or less than in our Universe.

In General Binary Physics, there are four factors that affect change:

(1) The ratio between the Existence and Placeholder in the Initial State. For example, in a universe that contains only the Placeholder, we cannot speak of change or potential change (assuming, of course, that the conservation rules—e.g., the Fourth Rule—are fulfilled also in the universe under discussion).

(2) The Existence Algorithm. For example, in a universe in which the Fourth Rule is not fulfilled, and one Existence can change over in the next turn to two Existences or 100 Existences—of course there could be potential change infinitely greater than what is familiar to us—this will be a universe that is many times more energetic.

(3) The number of connections of each cell to other cells. In the chapter on the Binary Field, we saw that the more connections per cell, the greater the respective number of possible dimensions in space and in depth. As each cell will have more connections, so will the transition from potential change to actual change be more frequent (that is, the transition from mass to energy). In a universe possessing many more connections (possessing more dimensions), it will be more difficult to preserve the obstructions, because each "Existence" will have many more opportunities to "escape" from the obstruction. The transition from potential change (mass) to change (energy) in a universe like this is expected to be much quicker. Therefore, it is reasonable to assume that in a universe such as this, we will see many more incidents of explosions (mass that explodes on its own) and the systems with the internal obstructions (the substances) will be much less stable.

Alternatively, in a parallel universe with fewer connections per cell, for example, a universe with only two spatial dimensions, it would be much easier to produce obstructions, and for the Existence it would be much more difficult to "escape" from the obstruction. In such a universe, the systems with the internal obstructions (substances), are expected to be much more stable. The transition from internal obstruction (mass) and change (energy) in such a universe is expected to be much slower. This universe will be much less energetic.

(4) The cell's response time.

In our Private Physics, the Fifth Rule determines the uniformity of the queue, and hence the uniform response time of all the cells. Thus, in our Universe, where the response time of each cell is uniform, response time has no effect on the potential change. However, General Binary Physics allows for the existence of different response times for the cells.

For example, in a universe with a faster response time, a faster transition between the potential change to actual change—and thus a wider spectrum of energies between a state of total obstruction to a state of free flow—will be possible. The speed of time in such a universe will be higher than in our Universe (I will expand on the concept of speed of time in the chapter on the illusion of speed). An observer from our Universe looking at such a universe will see processes occurring faster than in our Universe. Of course, if the response time of the cell would be higher in that universe, then also the maximal possible "speed of light", which constitutes free flow, will be higher than the "speed of light" in our Universe. The fastest processes in the parallel universe, such as

free flow of the Existence at "the speed of light", an observer in our Universe cannot grasp except at low resolution, because these are processes that are faster than the fastest possible process in our Universe.

It is also possible that a universe with different response times for different cells exists. In such a universe, areas where there are cells with a faster response time would allow for more energetic states than areas in which there are cells with a slower response time.

An extreme state is a universe in which the cell response time tends to infinity. In such a universe, change will not be possible, and the potential change will tend to the maximum. In other words, this will be a universe with energy levels that tend to zero.

And as a side note: If we have zero response time, the potential change will tend to zero and the change will be maximized. This is a universe possessing a tremendous energy.

The Photon

According to Classical Physics, the photon is the particle that composes the electromagnetic wave. It is a particle possessing energy but lacking mass.

Indeed, according to Binary Physics, the pattern in depth of the electromagnetic wave does not contain any obstructions, and therefore it does not have mass.

The pattern of a point in depth of the photon will appear like this (an abstract example for illustration):

1000100010001

The understanding that energy and mass are simply two different patterns of the Existence in depth solves the problem for us of the wave–particle duality of the photon.

Although the photon has no mass (because it has no obstructions in depth), certainly it has energy. After all, every free Existence contains one unit of change = one unit of energy.

Therefore, it is clear why when an electromagnetic wave in the right frequency hits an atom, we see Einstein's famous photoelectric effect:

The photoelectric effect occurs because the system of the photon in depth contains free Existence, which transfers the direction of its movement to the atom—according to the Ninth Rule of Private Binary Physics—when it "strikes" another Existence of an atom. Therefore, an electromagnetic wave that strikes metal agitates its system in space and in depth, and causes electrons to jump out of it.

The photon is definitely a particle—which consists of free Existence—with each one possessing one unit of change.

On the other hand, it exhibits features of a wave, such as diffraction, dispersion, and interference (as proven in the double-slit experiment of Thomas Young). The wavelike features are an expression of the fact that the electromagnetic wave consists of a cyclical free (massless) flow of the Existence in depth.

How is the phenomenon of diffraction created?

When the stream of the Existence in depth encounters a sufficiently thin obstruction in space, the Existences in depth that comprise the electromagnetic wave start to strike the obstruction one after another. According to the Tenth Rule—Gravity (Bypassing in Depth) —the Existence will try to bypass the obstruction in a circular motion. After it bypasses that obstruction, it will continue in its motion in the same direction in accordance with the Eighth Rule, the Rule of Inertia. Thus we will see part of the Existence, which strikes the edges of the obstruction in depth, changes its direction in space, and also reach areas beyond the obstruction – regions that should not be reached according to geometric optics (where rays of light move as straight lines).

How is the phenomenon of interference created?

Interference is actually an integration of two systems of an electromagnetic wave in depth.

Of course, we can see the phenomenon of interference only at low resolution. At the fundamental resolution, there is no interference, since every Existence is an independent unit.

When we look at the Universe at low resolution, we actually see a number of Existences as one unit. Therefore, when two streams of the Existence in depth (two electromagnetic waves) join together, we occasionally see an increase in the frequency of the Existence in depth (that is, constructive interference). We interpret this as a more energetic wave that is created in space, and sometimes due to the nature of the patterns in depth, when two areas of the Placeholder in depth merge, we see a decrease in the frequency of the Existence in depth (that is, destructive interference), and we interpret this as a less energetic wave.

The definition of the photon, according to Binary Physics:

Photon – "A system of a free Existence in depth"

Classical Physics defines the Planck constant as the constant of proportionality between the energy and the frequency of the electromagnetic wave.

Binary Physics declares the symmetry of space–depth. This symmetry is derived from the Sixth Rule of Private Binary Physics, which states that any movement in space requires a movement in depth.

That is, the higher the frequency of change in space, the higher the frequency of change that is required in depth.

That is to say, an ultraviolet electromagnetic wave with a high frequency of change in space has to have a high frequency of change in depth as well.

In contrast to this, an infrared electromagnetic wave with a lower frequency of change in space has to have a lower frequency of change in depth as well.

Once the frequency rises above a certain level, the flow ceases to be free, and "blockages"—obstructions in depth—begin to form.

Once the frequency in depth passes the critical threshold and obstructions start to form, what we call "light" becomes what we call "particle".

It should be noted that a light beam is actually a series of systems of photons moving in one direction. Patterns of Existence in depth with no obstructions which move randomly in space do not create a beam of light.

Energy is change. As the frequency of the Existence in depth gets higher, change increases accordingly. A higher frequency in depth means more Existence that flows in depth. More Existence flowing in depth means greater change in space. In other words, there is a direct relationship between the increase in frequency in depth and change. Therefore, it can be determined that change increases as the frequency with depth increases. Since the Sixth Rule determines that the frequency in depth is symmetrical to the frequency in space, we can say that change increases as the frequency in space increases. In the language of Classical Physics: As the frequency of the electromagnetic wave in space increases, its energy level increases respectively.

This is the source of Planck constant—that is, in the language of Classical Physics, the relationship between frequency and energy.

Indeed, we see also in Binary Physics that any increase of a unit of frequency in depth is equal to an increase of one unit of frequency in space, which is equal to an increase of one unit of change (energy). And one increase in change equals what Classical Physics calls "a Planck". Go and learn:

Planck - "One unit of change"

$E = MC^2$

The private Existence Algorithm allows for an Existence to have two states in depth—free and obstructed (Existence that cannot move due to other Existences obstructing it in the neighboring cells, in accordance with the Third Rule—the Rule of Non-Mixing).

Free Existence = Change (Energy). This is indicated by the units **Exfr**
(A contraction of the beginnings of the words "Existence" + "free")

Existence that is obstructed within the system = mass = potential change (potential energy). This is indicated by the units **Exob**
(A contraction of the beginnings of the words "Existence" + "Obstruction").

The overall number of Existence in the system (obstructed + free) is indicated by the units **Ex**.

The maximal potential for change in a system = potential change (Obstructed Existence = the mass) + change (Free Existence)

According to the private Existence Algorithm (The Seventh, Eight, and Tenth Rules), if Existence is not obstructed, it must move between the cells and create change. Therefore, any internal obstruction in the system (mass) that decays must convert to change (energy).

And conversely, any pattern of free Existence that forms an obstruction within the system – will convert change (energy) into obstruction (mass).

Change (energy) is a potential of obstruction (mass), just as obstruction is a potential of change.

The quantity of obstructions is measured by the number of times that Existence doesn't moves from its place in a given space-depth. Its units are **ob** (an abbreviation of the word "obstruction"). We must note the distinction between the quantity of obstructions and the quantity of Existence that is obstructed. These are two distinct concepts. One Existence can be obstructed a number of times in a given space-depth. Therefore, in general, the number of obstructions is greater than the number of obstructed Existences. The obstructed Existences are the mass, and not the number of obstructions.

Change (energy) is measured by the number of changes (substitutions of Placeholder-Existence) in a given space-depth. Its units are **ch** (an abbreviation of the word "change").

An example of calculating a system with pure change (energy):

The number of free Existences: 5 Exfr
The number of obstructed Existences: 0 Exob
Resolution: R(10)
The number of turns in depth, that the unit of time of the observer contains: 10 tu
The number of cells in space that contain the measured unit: 10 cl
(Explanation: The Sixth Rule (Symmetry of Space-Depth) requires that at resolution 10, any imaginary cell that contains 10 tu (10 cl in depth), will in essence be also a contraction of 10 cells in space.)

What is the change (energy) of the system?
5 Exfr*10*10 = 500 ch

That is, in one unit of time of the observer (which constitutes 10 objective turns of the system), 500 changes occurred in the system (substitutions of Existence-Placeholder in the cells).

Explanation: We can take, for example, a game in which in each turn, each piece moves 10 squares.
How many squares will 5 pieces moves in 10 turns?
500 squares.

This is of course assuming that no piece obstructs another piece.

The pieces are the Existence. The squares are cells. 10 squares that a piece moves in a turn represent the resolution of space, and 10 turns represent the resolution in depth.

At the Fundamental Resolution, the speed of the reaction time of the cell is one cell in one turn—1. For a free Existence, there is no possibility of moving at a lower speed than this. At the Fundamental Resolution, the maximum and only possible speed is 1.

The reaction speed of the imaginary cell (composed of two physical cells) at Resolution R(2) will be two cells in two turns. This is because, at R(2), the Existence must traverse two physical cells in order to traverse one imaginary cell. At this resolution, the maximal possible speed will be 2. But in Resolution R(2) another thing will happen. There will be another speed, a lower speed – 1. In R(2) we will see the speed of 1, in a situation in which the Existence enters one of the cells and remains there for one turn due to an obstruction in the second cell.

In the same manner, the maximum reaction speed of the imaginary cell at resolution R(10) will be - 10 (10 cells in 10 turns). Also, at resolution R(10), it will be possible to lower speeds to exist, such as the speed 1 (in the situation where there is a delay due to an obstruction in 9 cells) or a speed of 7 (in the situation where there is a delay due to an obstruction in 3 of the cells).

We see that in Binary Physics the maximal speed at any resolution is equal, in value, to the level of the resolution. We also see, that the amount of different possible lower speeds (levels of speed), is equal to the value of resolution. For example, in R(100) the maximal speed will be 100 (when there will be no obstructions in the system). And in R(100) there will be 100 possible levels of lower speeds (dependents on the amount of the obstructions in the system).

At the resolution of human systems R(h), the maximal speed in the state where there is no obstruction (there is pure energy) is c, the speed of light. If we translate this into Binary Physics, then the maximal possible speed is equal to the resolution, and thus R(c) = R(h). C is the maximal speed in which an imaginary particle can travel from one imaginary cell to another at this resolution. One imaginary cell at resolution R(h), is composed of a huge quantity of physical cells at the Fundamental Resolution. Therefore, due to obstructions (among the physical cells that contain the imaginary cell), many levels of speeds are possible that are lower than the maximal possible speed, but it is understood that there is no possibility of moving at a speed higher than this speed.

According to the Sixth Rule (Time Symmetry of Space-Depth), any movement of Existence in space requires that there also be a simultaneous movement in depth. Therefore, the number of turns that will transpire in the passage of the particle from one imaginary cell to another, must also be c.

At resolution R(h), the following formula results:
Change (Energy) =
The number of free Existences * c (the speed of reaction time of the imaginary cell in space in terms of R(h)) * c (the speed of reaction time of the imaginary cell in depth in terms of R(h))

And in short -
Change (energy) = Exfr*c2

Clarification: I say "one change in space in every turn", even though there are three spatial dimensions. This is because, according to our private Existence Algorithm, in every turn, Existence can advance one cell in depth and to advance only one cell in space (regardless of the direction). In one turn, there are not three changes in the spatial dimensions, but only one change.

An example of calculating a system that contains an obstruction (mass) alongside change (energy):

Total Number of Existences (free and obstructed): 6 Ex

Resolution: R(10) (that is, each imaginary cell contains 10 cells in space and 10 turns (cells) in depth)

Change (Energy) that is measured within the system = 200 ch

What is the obstruction (mass)—the number of obstructed Existences in the system?

First, we are required to calculate the maximal potential for change of the system if it were without obstructions (mass), that is, if there was pure change (pure energy). We will do this as we saw in the previous example.

6 Ex*10*10 = 600 ch

The change measured according to the data is 200 ch. Since every Existence that doesn't move, in any turn, is obstructed, the quantity of obstructions that were in the system is 400 ob.

In order to identify the number of Existences that were obstructed and constitute the mass, we must divide the number of obstructions by the number of cells in depth at the given resolution (10), and then to divide again by the number of turns (cells) in depth (10). Thus we will receive 4 Exob. That is, the obstruction (mass) of 4 Existences in the system. The mass is the number of obstructed Existences in the system (internal obstructions) at the given resolution.

We have seen that if we do not want to measure the quantity of obstructions directly, it is possible to calculate them from the difference between the maximum potential for change (that is derived from the quantity of changes in the system in the state in which there are no obstructions), and the quantity of changes that were actually measured.

That is, according to the number of changes (energy) and the number of Existences in the system, we can know what its mass is.

Calculate the complete conversion of an obstruction (mass) to a change (energy):

Obstruction (mass) = 4 Exob

Resolution = R(10) (that is, each imaginary cell contains 10 spatial cells and 10 cells in depth (10 turns)).

The number of turns in depth that contains the time unit of the observer: 10 tu

The number of cells that contains the measured unit: 10 cl

(Explanation: The Sixth Rule (Symmetry of Space-Depth) requires that at resolution 10, any imaginary cell that contains 10 tu (10 cl in depth), will in essence be also a contraction of 10 cells in space.)

Change (energy) = Obstruction (mass) * reaction time of the imaginary cell in space * the speed of the reaction time of the imaginary cell in depth

Change (energy) = Exob*10*10 = 400 ch

In fact, when the obstruction is removed, all of the obstructed Existences are converted to free Existence and start to move. At the Fundamental Resolution, every free Existence moves one cell per turn. At Resolution R(10), every Existence moves in each turn 10 cells, and in 10 turns, which are one imaginary turn at this resolution, it performs 100 changes.

As we have seen previously, at resolution R(h), the following formula results:

Change (energy) = Number of obstructed Existences * c (the speed of the reaction time of the imaginary cell in space in terms of R(h)) * c (the speed of the reaction time of the imaginary cell in depth in terms of R(h))

In short -
Change (energy) = Exob (mass) * c^2

And in fact, we have received the famous equation of Einstein:

$E = mc^2$

Change, Mass and π

Energy is change.

The Tenth Rule implies that change in space is circular. That is, energy is a circular concept. In the constant universe, energy and the circle are two identical concepts that are interrelated.

Therefore, although we observe the Universe at a very low resolution, we should not be surprised to reveal a similarity, if not an identity, between the structure of the formula for energy = change, and the formula for the circumference of a circle.

$E = mc^2$

Area of a circle = πr^2

That is, in the constant universe, at the fundamental resolution, change (energy) is the area of a circle.

In the constant universe, the response speed is a turn. That is, 1. That is c=1.

Due to the Sixth Rule (Symmetry of Space–Depth), the cumulative reaction time of a number of cells is identical to the number of cells under consideration. That is, the radius. Thus, r = 1.

From here we can learn something else: π is the basic value of one unit of mass = the basic obstruction. That is, the relationship between the obstructed Existence (the diameter) and the Existence that is obstructing them in the next depth (the circumference). The jumps in mass are jumps of pi. It is possible that we will discover that the circumference is the smallest obstruction that will prevent the diameter from decaying…

There must be a circular obstruction in depth of 3.1415… Existences in order to form an obstruction that will be interpreted as mass.

Precisely this obstruction is required, due to the geometric structure of our Private Physics.

As we decrease in resolution, physical reality blurs, but because π is such a fundamental number, we repeatedly find it again and again.

Of course, in the constant universe, the physical π is not the decimal number that we are used to seeing. The physical π is a geometric arrangement of the Existence in the Binary Field, a type of obstruction that is composed of three Existences in depth plus another partial Existence that stems from the fact that the obstruction can be spread statistically over a number of cells.

I will admit that this matter should be further investigated and refined. But if you are searching for π, proceed in this direction.

And if we are already speaking of an orientation to find pi, I will also give some direction of how to find the constant i (the square root of -1). We encounter i many times in calculations of magnetic fields that are formed by attraction in depth = spin in depth. Why?

This is because i constitutes the product of two numbers possessing the same value in space, but each one is fixed in a different level of depth. The different depth gives one of the values its negative value with respect to the other, and thus we receive a solution in four dimensions to what seems to be a paradox in three dimensions. The square root of -1 is composed of the product of two identical values, where one of them receives a negative value due to its different location in the dimension of depth, in relation to the first one.

An interesting thought from General Binary Physics:

Assuming that the response time of the cell is zero, then change (energy) of the system will be infinite.

Assuming the response time of the cell is infinite, then the change (energy) of the system will be zero.

Work

Classical Physics defines work thusly: "Displacement of mass as a result of exerting force in the direction of motion".

Let us convert this concept in light of what we have learned about the nature of the fundamental force in Binary Physics:

> *Work – "Removing a continuous obstruction in depth along a number of turns"*

When you begin to remove an external obstruction from a system along a particular depth axis, - the result will be a movement of the system in the spatial direction of that axis. The mental interpretation of the appearance that will be obtained in space will be that of a body moving in a certain direction as a result of a force that was exerted on it.

When I travel to ancient cities and I see castles, at times I believe with difficulty that what allowed these spectacular patterns of castles to develop in space is actually a removal of obstructions in depth, which we imagine, due to our blindness to depth, as work.

Chapter 20
The Birth of Substances and the Nature of Mass

As we have seen, according to the first axiom of General Binary Physics, only one particle exists, the "Existence" (the Placeholder is a Passive essence).

How can such a wide variety of what look like different materials, with different properties, can be formed of it? Or, to be more precise, how can human consciousness perceive such a wide variety of materials in observations?

From the perspective of physics, there are no other materials other than the "Existence" particle.

If we take a single cube of space and allow a human to contemplate the patterns of the "Existence" that is inside it, he will perceive only one type of material with a variable density level. For that matter, if we imagine that the Existence is a black square and the Placeholder is a white square, we could see in the spatial cube totally black pattern of the Existence that are adjacent to each other, or totally white patterns, which are areas of the Placeholder. At the moment when the human views the cube of space at a low resolution, he begins to see patterns in shades of gray. These will be patterns composed of a varying mixture of the Existence and Placeholder.

However, everything is "black and white" in the above example. There are no colors.

There is no explanation in the above example for different colors or different properties of materials. If everything is composed of the Existence only in Three Dimensions, how can we explain the fundamental properties of materials?, How can we account for things having a different specific gravity? How can we explain the reality of particles with different forces of attraction? How can we explain positive or negative charge? How can we explain different spin states?

In order to explain all these properties, we are required to integrate two fundamental concepts of Binary Physics, the depth and resolution.

Materials, according to Binary Physics, are formed **in depth (on the time axis),** not in space.

In order to understand this, we must let go of the very basic illusion that we learned from our childhood, "the present time".

Ask any child where he lives in terms of time. If he is clever, he will answer you immediately, in the present. If we are dealing with a very smart child, he will answer you the truth that he doesn't know.

We, human systems, treat the present as if this is one objective unit of time that contains a single spatial cube.

The emphasis is on one time unit and one spatial cube. We perceive the present as one time unit which contains all space seen by us at the same time.

In other words, our starting point is that if we "freeze" the present time, and a human system walks about in space, he would see all the materials exactly as we know them. We can see such scenes in movies.

The initial premise of classical physics is that all the materials exist as they are at present. If we take one slice of time, we would see the electron as it is, and the proton as it is. Perhaps they will be in a spatial position that would be difficult for us to predict, but their essence and their underlying properties will not change. In terms of classical physics, the past has no influence on the present (except a causal relationship), and the future certainly has no influence on the present. At the moment when the past has become the present, the past can no longer affect physical reality.

Binary Physics disagrees with this determination completely. From a physical standpoint, there is no such thing as "present" time.

Space as we perceive it does not exist in a single "unit of time". It is rather spread over many "time units", or, more accurately, over cubes of space in depth (the dimension of the time axis), so when put together they create materials and their properties as we know them.

If we take the minimal possible "time" slice — a turn —and consider an isolated spatial cube, there will be no "matter" in it. It will contain only the Existence. It will not contain neither protons, nor electrons, nor neutrons, nor light waves. Only the Existence.

But if we take 100 spatial cubes spread over a depth of 100 turns, and our brain interprets them all at once, as one cube, in which each cell consists of an average of 100 cells that constitute it in depth, we will suddenly receive an imaginary spatial cube where each cell has 2100 possibilities for different types of materials. Each material reflects a different pattern of the Existence and Placeholder in that cell's dimension of depth. If we compress a depth of millions of turns, we will receive a huge range of possibilities for different materials with various properties "in the present".

I will try to explain this by an example.

When we look at a television screen, we do not see pixels, but rather we see a picture, for example, an image of a car. This car has properties, color, length, width, different materials like glass and metal, *etc.*

If we look at the picture of the car at a higher resolution, we will see that in fact there is no car, but only pixels in three different colors. Each pixel by itself shares none of the properties of the car, not the length, not the width, and it does not appear like glass or metal. It may be that the car also has colors that are not one of the three primary colors of the pixels—this is because the instant that the picture is composed of many pixels, our consciousness visualize many colors that are in essence different mixtures of the primary colors.

As is familiar to us, when we look at a screen from a distance of two meters, we are viewing it at a low resolution, and our brain compresses thousands of pixels into a single image, which is seen on the screen.

This image does not, of course, really exist (since in reality, there are only pixels), and it exists only in our mind.

The picture that the brain imagines can have many features: length, width, color, etc. The features of this virtual image are derived from the mixture of pixels that compose it…

This reality, in which our brain compresses many pixels in space and displays them to us as an imaginary picture, is accepted by us as something obvious.

This is how the mind works. When we look at the sky, a huge amount of material is converted into a bright point in the sky, which we call a star. Each star has its own magnitudes of brightness and size.

In fact, there is no true difference between electrons and protons that compose one star or another. Our mind shows us different stars only because it perceives them at a very low resolution and compresses the entire physical reality of them into a kind of average in a single point.

Binary Physics states that exactly the same operation is performed by our brain with respect to the dimension of the depth ("time"). The brain considers the depth dimension at a low resolution and compresses many cells in the depth into a single imaginary cell with properties being a sort of average of the pattern of all the Existence and Placeholder in all the constituent cells.

We are blind to the depth (to the dimension of time). Therefore, we are unaware of the process that our brain performs when it compresses the depth and gives us an interpretation of different materials. In terms of our consciousness, which is blind to the depth and sees only space, there exist different materials with different properties in space.

Our present time is actually comprised of a vast quantity of "time" units that our brain, due to its limited processing ability, presents to us as an imaginary single unit, in which each point in space is actually an average of all the depth units that compose it.

People talk all the time about a possibility of traveling through time, but the truth is that our existence at every instant is spread over many "time units". We are stretched out from what we call the past through the present and into the future at the same time.

Each body in space has a depth in the "time dimension", and it is this that determines the properties of substances in it.

As we consider space at low resolution and see complex patterns in space as an isolated spatial point, precisely in the same way our brain sees depth dimension at low resolution, and interprets many cells in depth as one cell in space.

Let us consider a single cell in space and study its pattern on the depth axis. This dimension is called "d(4)" in Binary Physics. I call a cell such as this "a point in depth".

0000111100001111

I remind you that we are not looking at this cell in space. This is not a sequence of cells in space. This is a cell in one location in space, for example x_0, y_0, z_0, and we perceive it in depth along the length of many spatial cubes d(4)0, d(4)1, d(4)2, and so on, on what we call "the time axis". In the example above, we see the value of this cell over 16 turns.

Our brain views the depth at low resolution, so therefore it is not able to see each cell in depth.

The brain perceives, for example, 16 cells of depth—and interprets them as one cell in space. Of course the resolution in which the brain "sees" the depth is much lower, and therefore the number 16 is only for the purpose of illustration. The actual number is a huge: 1.343×10^{50} cells of depth are compressed into one second of our consciousness (of a "static" human system on the Earth's surface). The calculation is presented in Chapter 28.

You will probably immediately ask how could it be that the brain sees a depth of 16 cells and interprets them as an isolated cell in space? There can be 65,536 (= 2^{16}) different possibilities to see this cell, which will result from the different pattern of the Placeholder and Existence in depth.

Quite correct!

Our brain, which is blind to depth, is not able to show us the depth that is composed of the Existence, and therefore it interprets the different mixtures as different materials!!!

A good example of this is that of the "Flat Ones", flat people, who live in a two-dimensional universe, on the surface of a table. They can move freely in any direction

In a two-dimensional space. Of course, they themselves are also two-dimensional. At the center of the table there is a large bowl-shaped depression. Of course, the flat people cannot see the depth of this bowl-shape. Their brain is able to understand only a two-dimensional reality. So when they come to the area of the bowl, this region is interpreted as a different type of material, with a strange power of attraction toward a particular point in the center of the table (the point which is at the center of the bowl).

The observations indicate to them that if they reach a certain point on the table (which is at the edge of the depression) this attraction starts to increase towards the center of the table (which is the center of the bowl, and its deepest part). Therefore, in the world of the Flat Ones, there are two types of materials: the ordinary matter where they go most of the time, and the special material around the center of the table, that draws them to the exact point of its center.

This is exactly how the brain of human systems, which are blind to the depth, interprets the existence of depth as different types of materials.

Now I will provide examples of what the binary depth pattern of different spatial systems looks like, and I will explain why precisely this depth pattern creates these properties.

A pattern of a point in depth that appears as
00000000001000000000100000000001
will be interpreted as radio waves

A pattern of a point in depth that appears as
000000000100000000010000000001
will be interpreted as infrared light

A pattern of a point in depth that appears as
000000001000000001000000001
will be interpreted as visible light

A pattern of a point in depth that appears as
0000000100000000100000001
will be interpreted as ultraviolet light

A pattern of a point in depth that appears as
000000100000010000001
will be interpreted as X rays

A pattern of a point in depth that appears as
000001100000011
will be interpreted as an electron

A pattern of a point in depth that appears as
111100001111000
will be interpreted as a proton

And so on...

The expression for the pattern of a point of space in depth I will henceforth call in short: "**The depth pattern**".

Of course, in terms of the quantity of the Existence and Placeholder presented above, these are only illustrative examples. In practice, as a result of the low resolution at which the brain interprets the depth, there is a compression of a tremendous number of cells in depth. But this example illustrates nicely how a different pattern of the Existence in depth causes us to grasp the point of the space as if it contains "different matter".

The main assertion of Binary Physics is that the properties of matter/ Electromagnetic waves in space are determined by a certain pattern of Existence and Placeholder on the depth axis (the time axis) of that particular matter/ Electromagnetic waves.

What can we learn from this example?

1. As the frequency of the Existence in depth rises, our brain interprets it as a more energetic particle in space.

2. According to the seventh rule (about trivial motion in depth), every Existence aims to advance to the next level in depth. When there is an obstruction, when one Existence blocks another Existence in depth, it creates a type of "traffic jam" that slows the flow of the Existence through depth. This obstruction is interpreted by our consciousness as a particle with mass in space. When there is no such obstruction (i.e., every Existence is separated by a Placeholder, and therefore they can flow freely in depth), our consciousness interprets this as a particle with no mass (for example, a photon).

3. There is symmetry in the frequency of space in depth: the more the frequency of the Existence in depth increases, so the frequency that is reflected to us in space will increase. For example, the frequency of the Existence in red light in depth

is low, and therefore the frequency of red light in space is low. Even the wavelength of red light in space is symmetrical to its wavelength in depth. If in depth we see the Existence over every 100 cells, then also in space we see Existence over every 100 cells. Proceeding with this example, if the frequency of the Existence of ultraviolet light in depth is high, then the frequency of ultraviolet light in space is high. The energy of ultraviolet light is higher, since in depth it contains more Existence over the same number of cells than infrared light.

4. Light and matter are the same. The difference between them is reflected only in the frequency of the Existence in depth. The Existence with a very high frequency in depth that cause obstruction in the flow in the depth axis will be interpreted by our brain as a matter. The Existence with a low frequency in depth that don't cause obstruction in the flow in the depth axis will be interpreted by our brain as an " Electromagnetic wave".

The frequency of the Existence in depth = "the number of times the Existence appears in a point in depth at a given resolution"

The depth pattern determines properties of matter. The "material" that we see in space is a kind of illusion created by our brain, based on the pattern of the Existence in depth. I will provide some examples of how different patterns of the Existence in depth determines different properties of matter.

I will note that a "depth pattern" don't have to be composed from only one cell in space, since, if we are speaking of a lower resolution, it can be a number of cells in space. What is important is that in the eyes of an observer interpreting this information, we are speaking of an isolated point.

Why does the electron spin around the proton (what is electric charge)?

Classical physics tells us that the electron revolves around the proton because of an attraction. The proton has a positive charge and the electron has a negative charge. The negative charge is drawn to the positive charge. Sounds convincing…

What is that attractive force? Why does the electron have one type of charge and the proton another type? Why are these two specific types of charges attracted to each other? In my opinion, Classical Physics does not give an adequate answer to these questions.

Let us try to look for better answers by means of Binary Physics.

Binary Physics states that the properties of substances are determined in the dimension of depth (the time axis). Therefore, if we want to understand why the electron is attracted to the proton, we need to search for the answer in the pattern of the two particles in the dimension of depth (the time axis).

We can get a hint as to the structure of their patterns from the fact that we see the electron and proton in non-random interactions with each other.

> *The First Rule of Systems: The Rule of Cooperation:*
> *"When one system performs a non-random interaction with a second system, the survival of one of the systems is dependent on the second, or they depend on each other."*

From the above rule, it follows that when we see in observations that two systems perform an interaction with each other in a way that is non-random—the reason will be that the survivability of one system is dependent on the other system. For example, a system of fire will always perform an interaction with wood or another flammable material. A human system will always perform an interaction with systems of food. A system of a fish will always perform an interaction with a system of water. It is possible to give an infinite number of examples.

Since observations indicate to us that the system of the electron is not in a random interaction with the system of the proton, we can conclude that the system of the electron assists in preserving the system of the proton, or that the system of the proton assists in preserving the system of the electron, or that both systems assist in preserving each other.

Binary Physics explains how the system of the electron and proton each aids the survival of the other.

In order to see how the depth pattern of the electron system helps preserve the depth pattern of the proton system, and vice versa, we will abstractly describe their patterns along the depth axis:

- A cross-section of a first pattern in depth: 111100001111000 (represents a proton)

- A cross-section of a second pattern in depth: 000001100000011 (represents an electron)

We see that the pattern of the electron in depth is in opposite phase to the pattern of the proton. That is, wherever in depth the proton has a 1 (Existence), the electron has a 0 (Placeholder), and vice versa.

An "opposite" pattern of Existence and Placeholder in depth allows the two patterns to spin one around the other. The Tenth Rule of Private Binary Physics—the progenitor of circular motion—states that when Existence encounters an obstruction in depth, it will try to circumvent it in a rotational motion, in a fixed direction. The depth pattern of the electron that encounters the depth pattern of the proton starts to circumvent it in a circular motion.

Therefore, at the instant that the two systems, that contain Existence and Placeholder in opposite phase in their depth pattern, meet, they will start to spin one around the other.

In the specific case of the proton and the electron, the encounter of the two systems will not only cause them to spin one around the other, but also that the spin—the entrance and exit of the "Existence" of the electron into the areas of Placeholder of the proton—will cause them to be preserved.

How does the entrance and exit of the "Existence" of the electron into the regions of Placeholder of the proton help preserve the system of the proton?

Imagine a line of cars in a traffic jam on a straight road. At the initial state, the cars are arranged as follows: 4 cars, and then a space the size of 3 cars, and then again 4 cars, and then a space of 3 cars, and so on. This is a specific pattern of a traffic jam. You certainly understand that at the instant that the line of cars will start to move, the orderly pattern will start to collapse. That is, the pattern of the arrangement of 4 cars in a row and then 3 spaces, will change. Suddenly groups of 8 cars in a row, or that 2 cars will separate off and form a pattern of 2 cars and 2 spaces, and then another 2 cars… In order that the pattern that existed at the initial state will survive throughout the length of the advance of the line of cars, there must be something that will maintain it. There must be a pattern of another system, in close proximity.

For example, let us assume that alongside this highway, there is also a line of motorcycles. In the initial state, the motorcycles are located in opposite phase to that of the cars. That is, opposite every space between cars, one motorcycle is located. When the line of cars starts to move, the motorcycles enter and exit in a circular fashion into the spaces between the cars, in such a way that they do not allow the cars to reduce the spaces between cars.

In such a situation, it is certainly possible to see the line of cars remain for a long time in the pattern that it had at the initial state. On the other hand, it is possible to say that when the line of motorcycles preserves the pattern of the line of cars, the line of cars also preserves the pattern of the line of motorcycles, and doesn't allow the motorcycles to join each other (because they have to spin around the cars).

The line of cars represents the proton's pattern along the depth axis (the time axis)

The column of motorcycles represents the electron's pattern along the depth axis (the time axis)

Of course, the case where the spin of the systems one around the other will also preserve their patterns, is a rare case (to illustrate by means of our example—that a column of motorcycles will be formed that will match in its phase exactly to the column of cars). Therefore, out of an inconceivable number of possible patterns of Existence and Placeholder in depth, we see only two fundamental systems that survived—one thanks to the other.

We call these systems by the names proton and electron.

Why does our mind see the opposite pattern as rotational motion?

I will explain by way of example. Consider a nut and bolt. The bolt will simulate the depth pattern of the proton, and the nut will simulate the depth pattern of the electron. The nut rotates around the bolt due to the fact that its spiraling threads occur in the opposite phase to the etchings on the bolt. In precisely the same fashion, the "Placeholder" of the electron is found to be in opposite phase to the "Existence" of the proton.

If we imagine the nut spinning around the bolt in two dimensions, and we are not able to see the spiraling threads and etchings, we would be able to say abstractly that there is an "attractive force" that causes the nut to move in a circular motion around the bolt. In two dimensions—we will require a very developed sense of imagination in order to describe to ourselves that the source of this attractive force is the physical pattern of the nut and bolt in three dimensions.

It seems to me that to explain the attractive force by means of the pattern of matter in depth is a more correct explanation than to say arbitrarily that there is matter with a negative electric charge and matter with a positive electric charge, and they are attracted to one another.

Electric Charge:

> *"A system possessing a depth pattern that allows systems with complementary depth patterns to spin around it"*

Returning to the above example:

Please note that there is also a difference between the quantity of the Existence in depth between the electron and the proton.

As we know from classical physics, there is a difference in "mass" between an electron and a proton. The mass of the proton is significantly greater than that of an electron.

How does Binary Physics explain the property of mass? After all, if everything consists of Existence, even light and matter, how does matter have mass and light does not? How does a proton have more mass than an electron?

My initial instinct (and perhaps that of the reader as well) is to link the mass, in terms of the amount of matter, to the amount of the Existence, in other words, to make mass an inherent property of the Existence. That is, to argue that all Existence has a unit of mass.

This explanation allows us to clarify why the pattern in depth of a Proton, in the above example, has a greater mass than the pattern in depth of an electron (because it has more Existence). But it does not help us to understand why the photon has no mass and the electron does. You can see in the examples above that the pattern in depth of photon also contains Existence. So if every Existence contains one unit of mass we will expect the photon will be with mass too. But we know the photon doesn't have mass.

Therefore, I realized that mass, as understood as a property familiar to us from classical physics, does not inherently stem from the Existence particle. It can be said that the Existence particle itself – is devoid of mass.

Mass is our interpretation of an obstruction in the dimension of depth.

I will explain.

The Sixth and Seventh Rules of Private Binary Physics require the movement of the Existence in depth.

When the Existence flows in depth without any obstruction, it advances at the maximum possible speed (which is derived from the speed of reaction of the cell). Such a wave of the Existence in depth will be interpreted by us as lacking mass, for example, a wave of visible light.

When the Existence encounters an obstruction in depth, clusters of the Existence are created ("traffic jams" of a sort), and the flow of the Existence slows down its advance in depth. This obstruction in depth, this traffic jam, is what is interpreted by us as mass.

As the frequency of the Existence in depth gets higher, the probability of forming an obstruction, which we feel and interpret as mass, rises up.

As the frequency of the Existence in depth increases, we receive in space a wave of higher energy. We will interpret this in space, as electromagnetic waves of increasing frequency. As the frequency in depth goes up, so will the energy of the wave in space does. At some point, the energy of the wave in space will reach such a level (due to the highest frequency in depth), that obstructions—a sort of "traffic jam"—will start to form in depth. These obstructions will be interpreted by our mind as mass.

The energy level of the wave in space increases => the frequency in depth (on the time axis) increases, according to the Sixth Rule (symmetry of depth in space) => obstructions ("traffic jam") in depth begin to form => the super-energetic electromagnetic wave in space starts to be interpreted as a particle with mass.

If we look at the example of the proton 111100001111000011110000, we will see that due to the great density of the Existence in depth many obstructions are formed and its progress in depth (that is, in the time plane) is slower and therefore we feel and interpret it as possessing mass. The great density of the proton in depth also makes it difficult for another wave of the Existence to pass through it. Therefore, when we stretch out our hand against a wall it is stopped. But, if we stretch out our hand toward sunlight, our hand will pass through with ease. The density in the depth of the light rays is much less, since the frequency of the Existence in depth of a light ray is lower.

The frequency of the Existence in depth, and consequently the matter's mass, also influences another central property of matter, its speed. I will treat this subject in a separate chapter. Presently, I would like to emphasize that the difference between all the waves that we refer to as electromagnetic waves and the waves, which we refer to as matter, is underpinned by the difference in frequencies of the Existence in depth. Electromagnetic waves have a low frequency of the Existence in depth, therefore no obstructions ("traffic jams") are formed in the wave and it flows freely in depth.

Conversely, the frequency of the Existence is higher in a wave of matter, therefore obstructions of the Existence are formed in it and its flow in depth is stopped. This phenomenon is interpreted by us as mass.

The higher the frequency in depth is, the greater the chance that an obstruction – a particle possessing mass - will be created.

All of the above prompt a simple definition of **mass:**

> *" A mental concept that reflects an internal obstruction in the depth dimension of the system"*

The definition of mass relates to an internal obstruction in the system. The internal obstruction in depth of the system is interpreted as mass, and this is distinct from an obstruction that is external to the system, which is interpreted as force.

The obstruction itself is the mass, not the Existence. The Existence that flows freely in depth will be interpreted by us as massless. The longer the obstruction in depth, the higher specific gravity of a particle is perceived by our brain.

> *Specific gravity: "The frequency and length of clusters of the Existence internal obstructions in depth, in a material system".*

If so, why is the pattern of the arrangement of Existence/Placeholder in depth of a proton based on this example: 111100001111000011110000?

According to the Seventh Rule—Trivial Motion in Depth—all Existence wants to advance to its next direct level in depth.

For purposes of illustration of our example, let's say that the depth axis (the time axis) extends from left to right, in such a way that every Existence "wants" to advance in every turn one cell to the right.

Within the depth pattern of the proton, we see a sequence of 1111. This is a sequence that constitutes an obstruction. The reason that this constitutes an obstruction is due to the Third Rule—The Rule of Non-Merging. Existence can only advance to a cell that contains Placeholder. In our example, the cell to which it is supposed to advance contains an "Existence"—therefore it is obstructed.

Such a pattern will be interpreted by us, who are blind to the depth, as mass.

For sake of comparison, we described previously the pattern of the arrangement of Existence/Placeholder in depth of visible light (a photon) thusly: 000000001000000001000000001

The differences are immediately apparent.

The frequency of Existence in depth of visible light is significantly lower. In our abstract example, there are three instances of Existence in the photon, as opposed to 12 in the proton.

Since the proton is more energetic than the photon of visible light, its frequency of Existence in depth is much higher.

A second difference is that in the depth pattern of the photon there are no obstructions—the flow of Existence is smooth. When in the next turn the Existence will "want" to advance a cell to the right in depth—it will have no difficulty in doing so. The next cell contains Placeholder. Thus the pattern of the photon can advance in depth (on the time axis), and as a derivative of this, it can advance also in space, at the maximal speed, and it will be massless.

The pattern of the proton contains obstructions that delay the advance of Existence in depth (on the time axis), and thus its speed in space also will be lower.

The electron is described as having a depth pattern of Existence/Placeholder arranged in such a manner:

000001100000011000000011

Since the electron has less mass than a proton, the obstructions that form its mass in depth will be fewer. In our abstract example, the sequence of obstructions is 2, as opposed to a sequence of 4 obstructions. Understandably, since the electron has mass, its depth pattern must contain obstructions, albeit smaller ones...

The speed of the electron in space is faster than that of the proton, since the obstruction that blocks its pattern in depth (on the time axis) is smaller.

Since the energy level of the electron is higher than that of the photon, but lower than that of the proton, the frequency of Existence in the depth pattern of the electron is higher than that of the photon (6 as opposed to 3), but lower than that of the proton (4 as opposed to 12).

It is understood that I chose smaller, simpler numbers, for purposes of illustration, and for simplicity's sake, I was not careful to maintain the true proportions.

Why Do We See Pairs of Protons and Electrons (atoms) so Frequently in Observations?

A pattern of a point in depth of a proton: 1111000011110000111100

A pattern of a point in depth of an electron: 000001100000011000001

In practice, the patterns are significantly more complex.

As we have seen, the pattern of the electron in depth is the complementary to the pattern of the proton.

We see that in the place where there is Placeholder in the proton's pattern, there is "Existence" in the electron's pattern, and vice versa.

The depth pattern of the proton is what allows the electron to spin around it, by the rule of the Tenth Rule. The "spacious" Existences of the electron try to circumvent the obstruction of the "dense" obstructions of the proton's Existence, by-passing in a circular motion. The "Existence" of the electron alternately enters and exits in depth into the regions of "Placeholder" of the proton. These patterns complement each other. We interpret this motion in depth, by superficially looking at space (without depth), as an attraction.

There is another significant insight, which I would like to draw your attention to. The spinning of the electron around the proton not only maintains the patterns of both in depth, but also, prevents an extraneous Existence from entering into the proton (and the electron) pattern, and in this way ensures preservation of the entire system.

Practically speaking, there are two factors that can cause "destruction" of the pattern in depth:

1. Internal collapse: the failure to maintain the Existence–Placeholder ratio in the pattern. The Existence begins to crowd into the pattern, and in this way increases its depth frequency, or, alternatively, the Existence starts to spread out in depth, and in this way its depth frequency decreases. This way or another, any such change will immediately influence the properties of the material: mass, charge, *etc.*

2. External interference: an Existence goes out from the pattern,

or enters in. The Existence leaving the pattern lowers its depth frequency, and entry of the Existence into the pattern increases its depth frequency. Any change in the Existence frequency in depth immediately affects the properties of the material.

The electron and proton are 'symbiotic' patterns, which, for the purpose of their survival, require one another, because their complementary patterns allow them to maintain the patterns of each other.

The electromagnetic force —

> *"The change that is created in the movement of the Existence as a result of spin in depth of a system possessing one pattern, around a system possessing a complementary pattern"*

In fact, any pattern that we see in space is composed of a much larger pattern in depth. The pattern that is hidden from the eye, but is found in depth, can create in a particular arrangement of Existence and Placeholder, an electromagnetic field in space. There are systems for which the part of their pattern that is found in depth significantly influences other patterns, these are systems with a significant electromagnetic field. There are other systems that their patterns are more "closed", and because of that patterns can't or almost can't spin around them. These systems have a weak or non-existent electromagnetic field.

Any change in the proton's depth pattern (a combination of the Existence, removal of the Existence, or change in frequency in depth) will directly influence the properties of the proton, and in effect from that moment on it will no longer be a proton, but another particle.

A common change such as this is the conversion of a proton into a neutron. This occurs when the physical pattern of the proton merges with the physical pattern of the electron (of course I am talking in abstract terms...).

Pattern of point in depth of proton - 1111000011110000111110000
Pattern of point in depth of electron - 0000011000001100000110
Pattern of point in depth of neutron – 1111011011110110111110110

What do we see here?

The pattern of the neutron in depth is actually a combination of the pattern of the proton with the pattern of the electron. Its mass is greater than the mass of the proton and certainly more than the mass of the electron (we see more significant obstructions in depth).

The neutron is neutral with respect to its electric charge. Its physical pattern in depth doesn't allow the electron to revolve around it, and due to its physical pattern it itself certainly cannot revolve around a proton or electron or around other neutron.

A depth pattern that does not allow other patterns to spin around is electrically neutral.

We also find neutrons inside atomic nuclei, because they also have to maintain their depth pattern. Neutrons being surrounded by protons, and the latter being surrounded by electrons, form a stable pattern in depth, in which the Existence does not get released.

Once the neutron goes out of the atom, with no electron that revolves around and maintains it, it starts to decay in a manner in which part of the Existence leaves it and its constituent electron, and it reverts back to being a proton. For accuracy's sake, more Existence is emitted in the process, forming an anti-neutrino.

I must make a note about the pattern of the proton in particular and the patterns of particles in general. Unlike my examples above, that were given for simplicity of explanation, the patterns of the particles and the proton in depth is not only pattern of a point in depth, but they also have a spatial distribution. For example, it may be that the proton is composed of a combination in depth of a number of points in space…

For example, the proton is composed of a number of patterns in depth, which together form its structure. Every one of these patterns is known in Classical Physics by the name "quark". These patterns are found one next to the other as well, because they possess depth patterns that assist them in preserving one another.

The different depth pattern of every quark is, what causes it to have a different charge (A reminder that charge = The ability of a pattern to allow other patterns to spin around it, in accordance with the Tenth Rule). Quarks have a charge of 2/3 or -1/3 of the charge of the electron. That is, a pattern that does not possess an opposite phase to the pattern of the electron. Therefore, the electron is not able to spin around an isolated quark. As is known, the quarks can be arranged only in a way in which they form together a whole multiple of the charge of the electron—that is, they form together a depth pattern which will allow the electron to spin around it.

Regarding the spin of the quarks—I will soon explain the concept of spin in what follows.

The Strong Nuclear Force

According to Classical Physics, the Strong Force joins the quarks. This force is a short-range force—on the order of magnitude of the atomic nucleus.

As we have learned, in Binary Physics, there is only one force—the force of the obstruction. This force is an external obstruction (as opposed to mass, which is an obstruction that is internal to the system).

The force that joins the quarks to one another is the obstruction, that results from their depth patterns, and which causes them to spin (in depth) around one another.

The fact that most of the structure of the system of quarks is found in depth, explains also why Classical Physics claims that approximately 99% of the mass of the proton does not result from the rest mass of the constituent quarks, but rather from the energy of the Strong Force. Binary Physics explains that because most of the structure Of the system of the quarks is located in depth in its pattern of obstructions (which form, what the mind calls "the Strong Force") that is also where its mass is to be found (the internal arrangement of obstructions in the system).

Why are precisely the electron and the proton the most common systems in our Universe?

It is very reasonable to assume that at the initial stage, and shortly afterwards, many types of flow patterns in depth developed, in addition to the patterns of the proton and electron.

Under any Existence Algorithm, there are patterns that are more suited for the algorithm, so these patterns have better ability to maintain their system than others.

We saw in Chapter 5 that the Existence Algorithm essentially creates the collection of private rules of a specific universe, which determine how the Placeholder changes to the Existence, and *vice versa*.

I will give an example.

Let's say that we have a two-dimensional universe. The Existence Algorithm of that universe determines that if there is an Existence to your right, you will always travel in its direction. If there is Existence to your left, you will always move to the left. If there is no Existence either to your right or left, then you will randomly move forward or backward.

In the initial state of this universe, there are two patterns:

Pattern A 01010

Pattern B 11111

Which pattern will survive for a longer period of time?

The intention here in "surviving" is that we will see the same pattern after an x number of turns.

I would place my bet on pattern B.

Pattern B is more suited for the specific nature of the Existence Algorithm that operates in its universe.

That is to say, if Pattern A and Pattern B were initially created by chance—over many turns—we would see that Pattern B would survive. On a larger scale, after enough turns, we would see a "universe" in which there are many patterns of Type B, and very few patterns, or none at all, of Type A.

It is possible to deduce from the above example an insight that in our Universe as well, there are patterns that are more "resistant" and better adapted to the character and specific rules of the Existence Algorithm of our own Universe.

Of all the many patterns that exist at depth, two have proven, and still prove, to be exceptionally resistant to the specific character of our Existence Algorithm.

These two patterns are the electron and the proton.

Because our brain interprets the flow pattern of the Existence in depth as a particle with different properties, it is reasonable to assume that during the first days of our Universe, there were millions of particles with different properties. From amongst them remained and developed two central patterns in depth, *viz.* the pattern of the electron and the pattern of the proton. They also developed in such a way that they are required to be next to each other in order to survive.

It should be noted that at each resolution at which we look at the universe, we will see different types of patterns. At the resolution at which we human systems look at it, we see the proton and the electron as two common patterns of what we call matter.

As we know, there are many more common patterns in depth, which have no obstruction in depth, which we interpret as light waves in their entire spectrum.

We saw earlier that the depth pattern of light waves is a more spacious format, without obstructions in depth, in which the frequency of the Existence is significantly lower that the frequency of the Existence in a pattern that we call matter.

The pattern that we refer to as matter contains obstructions in depth, which we interpret as mass, while the pattern that we interpret as light waves does not contain obstructions in depth, and therefore it can move freely and at its maximal speed.

The more complex the pattern is, with a higher frequency in depth and more obstructions in depth, the less chance it has to preserve itself through the turns.

The opposite is also true. The simpler the pattern, with a lower frequency in depth and the fewer to no obstructions, the greater chance does it have to maintain itself. This is true even without the need of cooperation of another pattern that spins around it.

Therefore, we see a wide spectrum of low-frequency depth patterns, which we call light waves, and a very small number of high-frequency depth patterns that possess obstructions, which we call matter (protons, electrons, and neutrons...). High-frequency depth systems, containing obstructions, require a complementary pattern that will spin around and maintain their pattern.

What is the difference between a particle and an anti-particle?

I will explain by way of example.

The pattern of a point in depth of the proton - 111100001111000011110000

Pattern of the electron – 000001100000011000000110

Pattern of a point in depth of the anti-proton – 000011110000111100001111

Note: These are abstract examples, for the sake of visualization only.

What do we observe here?

We see that the masses of the proton and anti-proton are identical. This is because the obstruction in depth of both of them is the same.

The frequency in depth of the two particles is the same, therefore the strength of their charge—the ability of other patterns to spin around them—is the same.

So what differentiates between them?

We see that the arrangement of the Existence and the Placeholder in the two patterns is reversed. Where there is "Existence" in the proton, in the anti-proton there is Placeholder, and vice-versa. Their phase in depth is opposite. The spin of the electron around the proton is performed in such a way that the "Existence" of the electron alternately enter and exit into the regions of the proton that contain Placeholder. Thus, if the depth pattern of the proton is in opposite phase, the "Existence" of the electron can't "enter" now, and cannot spin around it. We, who are blind to depth, will interpret this lack of ability of the electron to spin around the anti-proton in depth as a lack of attraction between them.

Since the pattern of the electron in depth doesn't allow it to spin around the anti-proton, the anti-proton will constitute an obstruction that will repel the electron in its attempt to go around it. We will interpret this in human resolution as a repulsive force.

Of course, the pattern of the anti-electron, whose charge is opposite to that of the electron, would match exactly the pattern of the anti-proton, and the anti-electron would spin around the anti-proton as if it would try to circumvent the obstruction of the latter. So if an anti-electron and anti-proton meet, we would see the formation of an anti-atom. Since the phase of both of them is opposite to that of the proton and electron, again, the "Existence" of the anti-electron is found opposite the "Placeholder" of the anti-proton, and they can spin around one another.

The Existence and Placeholder are symmetrical to each other. It is possible to view the Placeholder as Existence and *vice versa*. Therefore, one can say that the antiparticle is a reversed pattern in depth of the original particle—in terms of the composition of the Existence–Placeholder.

How does a particle change into an antiparticle?

When there is a tremendous energy burst in space (when we remove an obstruction and a great potential for change is realized), the particles that collide with each other affect not only their movement in space, but also their movement in depth.

Blocking the Progress of the pattern of a system in depth along several of turns, can change system into what we call anti-system.

Here are three identical systems:
The pattern of a point in depth of system A 001 100 110 011
The pattern of a point in depth of system B 001 100 110 011
The pattern of a point in depth of system C 001 100 110 011

As a result of removing the obstruction in space, a great potential for change is realized, that influences both space and depth. System B encounters the obstruction in depth of two turns, at the time that systems A and C continued to proceed in depth. After the change, the pattern of the three systems will appear as follows:

The pattern of a point in depth of system A 001 100 110 011

The pattern of a point in depth of system B 110 011 001 100 (a system that was blocked in depth)

The pattern of a point in depth of system C 001 100 110 011

If systems A and C are common depth patterns of substances and Pattern B is an unusual depth pattern, then human systems that are composed of the Type A (or C, which is the same thing) pattern will call the unusual matter that is formed from pattern B as anti-matter.

Note: To change an electron to a positron, the obstruction/collision must be in depth, and it is not sufficient to just have a collision in space. The lack of compatibility that needs to be created in order to change a property such as charge must be in depth.

As might be that following a collision in depth, the pattern of the electron will go out of its compatibility in depth and change into a positron, so also can it be that due to the same collision in depth a proton will go out of its compatibility in depth and change into an anti-proton. If this would happen simultaneously, we would receive anti-matter: that is, a positron that spins around an anti-proton. Of course, in terms of compatibility of the algorithmic environment, which is mostly composed of the proton and electron patterns, this would be odd to the anti-matter pattern, and its life span is expected to be short.

How do the electron and positron annihilate each other into photons?

00000011000000110000011 - electron (pattern of point in depth)
00110000001100000110000 - positron (pattern of point in depth).
00110011001100110011 - integrated pattern of an electron and a positron (pattern of point in depth). This is an unstable pattern, because there is no complementary pattern that will preserve its obstructions from decay.
001010101010101010101010 - Photon (pattern of point in depth). Obstructions of the integrated pattern decay and a photon is created.

The meeting of the electron and positron in depth and the integration of their patterns make them change into a pattern of an electromagnetic wave, which, in terms of its physical pattern in depth, cannot spin around the proton or around the anti-proton. Because the integrated pattern of the electron and the positron has no complementary pattern that will preserve it and its obstructions in depth, the obstructions decay. Thus a pattern—the massless photon—is created without any obstructions. (A reminder: the mass is our interpretation of an internal obstruction in the system.)

In all these examples, we must remember that in practice all the properties that we see exist only at low resolution, and we observe everything within a "compression" of depth to our resolution. That is to say, that in order to change a charge of an electron and form an anti-electron, it is not sufficient to have a delay of an individual turn, but practically speaking it rather requires a more significant delay. Due to the low resolution, a delay of a single turn will not suffice in order to reverse the phase of the pattern that create the charge.

Why are observations of antiparticles significantly rarer than observations of particles?

First of all, as we saw, both the particle and antiparticle are mainly identical, and consist of an Existence and Placeholder arrangement in depth.

The antiparticle is characterized by the same pattern as that of the particle, just in an inverse phase (mirror image of the pattern of the Existence and Placeholder in depth). We interpret this as an opposite charge (since the charge is the expression of the particle pattern in depth and the ability of this pattern rotate around other patterns).

In a specific environment, in which a particular Existence Algorithm operates, certain patterns are able to survive better than others.

In order for a particular pattern to survive, there must be a certain algorithmic process that will allow this and prevent the pattern from scattering and decaying.

Depending on initial conditions and the specific Existence Algorithm of that universe, it will be determined which patterns will succeed in being preserved and surviving.

In Private Binary Physics, in our universe, this algorithmic process is the electric charge. The electric charge is essentially the algorithm which causes one pattern in depth (the electron, for example) to spin around a second pattern (the proton, for example), and thus the patterns maintain one another from decaying. The two patterns are complementary patterns. The meaning of complementary patterns is that they preserve the obstructions from decaying (by entering into the spaces between the alternate obstructions) and prevent the Existence from "leaking out" from the pattern (by means of its own obstruction).

Reminder: Private Binary Physics is Binary Physics of our own universe. Within the framework of general Binary Physics, it is possible to describe a very wide variety of Existence Algorithms, where each one creates a new physics. In General Binary Physics, all possible universes are described, that include all possible Existence Algorithms, combined with all possible initial states.

In General Binary Physics, there may be many universes that, due to a specific initial state or a particular Existence Algorithm, would only allow for a chaotic reality, and would not allow patterns to develop and survive over time.

In our Private Binary Physics, we call the stable depth pattern by the name "particle".

Under certain circumstances when a different pattern is created, which does not have the complementary pattern to preserve it from decaying, or if it does not have an algorithmic process that keeps it stable, this pattern will decay.

In the initial state, at the moment of the Big Bang, there were no particles. When an observer external to the system looks at the distribution of the Existence and Placeholder in the universe at the initial state, he would not be able to determine where there are patterns and which patterns are stable. In order to know this, he would have to continue to observe the universe through a number of turns, to examine the way in which the Existence Algorithm influences the movement of the Existence, and to see which patterns, out of the variety that developed randomly, survive, and what is their System Algorithm (the physical rules pertain to them).

As viewers internal to the system, who cannot directly see the initial state, we must try to analyze why certain patterns are more common than others.

From the observations that we have before us, we recognize the Tenth Rule—gravity—which describes the principles of circumventing in depth. A derivative of this rule states that some depth patterns have a greater survivability than others. If the physical rule determines that one depth pattern tries to circumvent another by spinning around it, and the spinning is what maintains the stability of the depth pattern of the two patterns together, then systems with patterns suitable for such spinning will survive for longer than others.

It is reasonable to assume that in the initial state, there was a slightly higher incidence of depth patterns that we call particles. Patterns of particles started to spin one around the other, according to the Tenth Rule, and in this way one prevented the other from decaying. It is understood that the moment that a model had been created in which certain patterns preserved their stability, those patterns received an evolutionary survival advantage over other patterns.

Likewise, patterns of antiparticles of opposite phase also started to spin one around the other.

Occasionally, as a result of random obstructions in depth, a pattern of a particle turned into an antiparticle, and *vice versa*. Because the prevalence of particles had a slightly higher probability, the chance that an antiparticle that changed to a particle would find a complementary pattern and survive was higher than the likelihood that a particle that changed to an antiparticle would find a complementary pattern.

This process accelerated itself until the universe remained almost exclusively with patterns of particles only, and patterns of antiparticles became rare.

Of course, even at present due to a random obstruction in depth, a particle changes into an antiparticle. But since the present algorithmic environment is "unfriendly" for antiparticles and they are not able to find a complementary pattern to preserve them, the chances that their depth pattern will not be preserved and it will decay are high.

The depth pattern of particles and antiparticles is identical, except for their opposite phase. Our own Existence Algorithm has no direct effect on the survival of one type of the pattern or another, being dependent on the phase of the pattern in depth. The initial state is what had determined that under our specific Existence Algorithm we became a universe of particles. It is simple to describe a slightly different initial state under the same Existence Algorithm in which our universe would be a universe of antiparticles.

Of course, the definition of a particle and antiparticle is arbitrary. The particles in the common phase simply are referred to as particles, while those in the rare phase are called antiparticles.

A Cyclical Universe - Anti-Universe

If our consciousness could be delayed upon several turns in depth, we would see exactly the same universe, but composed of antiparticles. From the vantage point of our consciousness, it wouldn't even notice the change.

It could be said that the universe is built in depth of a layer of particles, and then a layer of antiparticles, and then again a layer of particles, and so on. This is a result of the pattern of particles in depth.

Particle 11001100
Antiparticle 00110011

In the example above, we see that if the consciousness would delay upon two turns, it would percept the antiparticle as a particle.

The consciousness of a creature internal to the system is not able to distinguish between a particle and antiparticle, rather only in a relative manner. These two concepts are relative ones. Because the essence of the difference between the particle and antiparticle is due to their opposite phase in depth, their definition stems from the point in depth from which an observer observes them. I refer to a particle as a particle only because most of the depth patterns that I see at a certain point are identical to it, whereas patterns of the opposite phase are rare. If I would look a few turns back in the universe (the number of turns equal to half of the wavelength of the pattern in depth), I would see exactly the same universe, only composed of antiparticles.

The depth pattern of the particles is cyclical. This creates layers of particle–antiparticle in depth.

Synchronization of systems with other systems in depth is critical for maintaining their pattern. The moment that a system leaves it synchronization and changes its phase with relation to other particles, it decays, because the entire preservation of systems depends on the interaction between them.

There is nothing that prevents, either axiomatically or due to the rules of Private Binary Physics, the existence of systems possessing opposite phase. Once a system of an opposite phase has been created—for example, an anti-proton—its chance of finding an anti-electron to spin around it and maintain its depth pattern, is low. Therefore, the life time of antiparticles is short as compared to a particle.

Why do antiparticles appear in particle accelerators?

Particle accelerators generate movement at a very high "speed". I write "speed" in quotes, because we will see later that this is a concept of our consciousness. What particle accelerators do is remove the obstructions in depth, and thereby they allow the Existence to move at its maximal speed, which is based on the response time of the cell. Removing these obstructions in a chain reaction allows for a very large potential change to be realized. This means a very large movement of the Existence. This large movement of Existence causes, among other things, a formation for an instant of new obstructions in depth. These obstructions cause the pattern of a certain particle to delay for a number of turns in depth, and to convert to an antiparticle (to change the phase of its depth pattern).

In terms of classical physics, it can be said that an antiparticle is a particle delayed in time, and thus it changed its phase.

What is spin?

Spin is defined in classical physics as an intrinsic angular momentum of particles. The principles of quantum mechanics determine that the observed values of angular momentum are limited to integer multiples of the Planck constant.

In Binary Physics, spin, like other properties of particles, is determined by the depth pattern of the particle.

> *Spin in Binary Physics is "Repetition in Depth".*

The meaning of spin is the wavelength in turns that is required in depth, so that the pattern of a system will repeat itself.

Whole spin can be contracted, and it looks identical at any resolution, like responsive Internet sites that are able to adapt themselves to any type of screen.

Since our means of observation, the brain of human systems, makes observations at low resolution, thus the concept of spin as repetition in depth becomes critical.

The repetitious nature of a pattern possessing half-integer spin is such that, at low resolution, half of the pattern contains the complete pattern in depth at high resolution.

A pattern possessing spin = 2 means that its repetitious nature is such that, at low resolution, it is necessary that its pattern advances twice in order that we will statistically be able to achieve full advance of the pattern in depth at the Fundamental Resolution.

Spin 1
At Resolution 1
0000111100001111
At Resolution 4
0101

By spinning the pattern at Resolution 4, we will receive the same result as its spinning at Resolution 1.

Spin 2
At Resolution 1
0011110000111100
At Resolution 4
1010

By spinning the pattern at Resolution 4, in half of the cases, we will encounter 0, while in half of the cases we will encounter 1. Only if we rotate the pattern twice shall we obtain statistically correct results.

1/2-Spin
At Resolution 1
0101010101010101
At Resolution 4
1010

By spinning the pattern at Resolution 4, it is sufficient to rotate it only 1/2 of the pattern in order to obtain the result. That is, by a complete rotation of the pattern, we will obtain the result of the pattern at Resolution 1, two times.

We connect through this explanation to the statistical statement of spin in Classical Physics: "Any orbital in an atom can accommodate a maximum of two electrons whose spins are opposite in relation to each other".

The reason for this is the nature of the movement of the electrons around the proton in depth, in which the patterns of the electrons also must not collide with each other.

In the initial state, it is reasonable to assume that many patterns of spins existed. Like in general, the patterns of the proton and the electron that survived the evolutionary process because their pattern was more suited to our private Existence Algorithm, similarly certain patterns of spin (reflective) survived the evolutionary process, because their character was more suited to our private Existence Algorithm.

What is a magnet?

> *Bar Magnet: "Obstruction with a linear gradient in depth, for which the Tenth Rule causes the Existence around it to flow in a certain direction"*

> *Horseshoe magnet: "Obstruction with a semicircular gradient in depth, for which the Tenth Rule causes the Existence around it to flow in a certain direction"*

Because the magnet is an obstruction of the gradient in depth that causes a flow of the Existence in a certain direction, thus when we bring another magnet close to it—which direction of the gradient is opposite—the two opposite currents of Existence in depth create what our consciousness interprets in space as repulsion.

When we bring two horseshoe magnets close to each other in the correct direction—their obstructions will complement the form of an ellipse in depth, and will form around this obstruction a circular spinning motion of Existence in depth—something which our consciousness grasps in space as attraction.

When we break a magnet in half, we get two magnets. Essentially what is done when we "break" a magnet is to split the obstruction in depth into two separate obstructions possessing a gradient in depth, so that each one individually will cause an identical effect in space.

When we connect two bar magnets together, we essentially create a longer obstruction of a gradient in depth.

Magnetic Field Lines

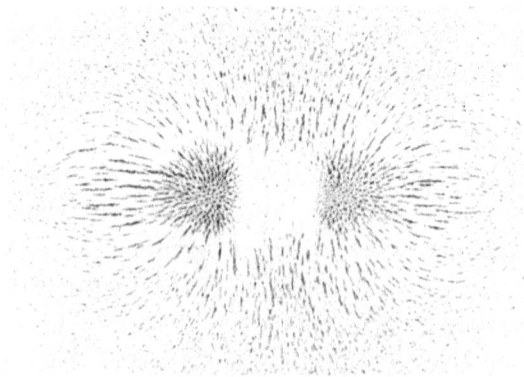

When we look at the picture of the magnetic force lines, we see a flattening of the flow in depth. The magnet is a graded obstruction in depth. The north pole of the magnet is an area with a "high" level of obstruction in the depth dimension d(4). The south pole is an area with a "low" level of obstruction in the depth dimension d(4). In order to circumvent the obstruction, a flow of the Existence is created from the north pole to the south pole. When our consciousness, which is blind to depth, interprets this flow in space, it sees it as unexplained force lines that are formed around a magnet. These force lines are essentially a flattening of the flow lines of the Existence in depth. The Existence "spills" from an area where the obstruction in depth is high to an area where the obstruction is low. In essence, our consciousness sees a three-dimensional picture of the four dimensional process that occurs in depth.

Dark Matter (Dark Energy)

Dark matter is in fact the Existence in depth that significantly influences the flow of the Existence in space.

The part of depth that our consciousness compresses and presents to us at low resolution as materials, which we are familiar with, is probably only a fraction of the depth that constitutes the flow of that material. The entire part of depth that our consciousness does not compress, but which has no less effect on the flow of the Existence in space, is dark matter/energy.

Basic Systems in Space

The main system of the particles—proton, electron, photon—is in depth.

The most basic system, which is familiar to us from observations and has a meaningful expression in space, is that of the hydrogen atom.

This system is composed of one proton in the nucleus and one electron that revolves around it.

According to Classical Physics, what binds a proton and electron is electrostatic attraction. The proton is positive and the electron is negative, and as a result of this arbitrary definition, they are attracted to each other.

Binary Physics says that the attraction between a proton and electron is due to their physical pattern in depth (the arrangement of the obstructions in depth) that allows them to spin around each other according to the Tenth Rule. The spinning in the depth of the system of the electron around the system of the proton also preserves the two systems from collapsing (that is, it maintains their stability).

The spinning of these two systems, which form long streams of the Existence in depth, is grasped by our mind "in a smoothed-out manner" in space as the spinning in space of one particle around a second particle.

Let us take a cross-section of a point in depth. The meaning of "a point in depth" is a point in space (cell) that we look at, in one dimension, the dimension of depth.

The system of the proton in depth looks more or less like this:
111100001111000011110

Of course, this is just an example to illustrate the fundamental system, and not meant to be at all accurate.

The system of the electron will look more or less like this:
000001100000011000000

First, we see the difference in masses between the electron and the proton

(The longer the sequence of the obstructions, the larger the mass.)

Second, we see patterns that complement each other in depth.

An electron that tries to circumvent the obstruction in depth that forms the proton will start to spin around it, according to the Tenth Rule.

The spinning maintains the system in two ways:

First, it prevents foreign Existence from entering into the system.

The spin of the pattern of the electron around the pattern of the proton, is performed in such a way that the Existence of one pattern alternately enters and exits the second pattern, and thus obstructs the entry from another foreign existence into it.

Second, it prevents the Existence within the system from going out. The spinning is such that, each time an Existence "wants" to leave the system, it will be blocked by the Existence that spins around it.

Go and learn: In order to create a spin between two particles, according to the Tenth Rule, the systems have to complement each other in depth, in such a way that spinning will preserve their stability and prevent them from collapsing, by preventing exit/entrance of own/exotic Existence out of/into the system.

How are more complex atoms formed?

More complex atoms are formed in exactly the same way and for the same reasons, which create a hydrogen atom, only their complexity increases in both space and depth.

If there is a more complex system that is composed of a greater pattern of the Existence in depth, which is distributed over space (i.e., a number of protons), it will require a larger array of small patterns in depth (electrons) that will spin around it, in order to keep this flow stable, and to prevent the internal Existence from going out, and foreign Existence from coming in.

The Tenth Rule, that creates rotational motion, cause spin of patterns of Existence around one another, and essentially allows for the connecting of systems to one another.

How are molecules formed?

Because systems developed evolutionary (the better the pattern of obstructions that composed the system suited the Existence Algorithm character, the more its survival chances increased), only certain systems succeeded in surviving. Thus, over the span of the turns, started to develop more and more sophisticated patterns, being made of obstructions and spinning, that preserved the systems in a state in which Existence continued to be dispersed.

Patterns in depth that started to spin around other patterns in depth and allowed them to remain stable, survived, while the patterns in depth that had not found another pattern to spin around them, did not survive.

After the basic stable system of the hydrogen atom, which was based on the spin of one pattern (the electron) around another pattern (the proton), had developed, other systems started to develop on the basis of this principle. Then heavier atoms started to appear, that was based on a number of patterns of electrons in depth that spun around a number of other patterns of protons in depth, and preserved them.

Subsequently, an even more complex pattern of obstructions started to develop, in which a one pattern of the Existence in depth of an electron spun around two systems of the Existence in depth of atoms, and thus allowed for their survival. In concepts of Classical Physics, it forced two atoms to combine into a molecule. I say "forced", because systems of atoms that lacked a single electron that would spin around them in depth and preserve their system did not survive. Therefore, we see only the systems that managed to survive in the environment of the private system algorithm.

Every universe has its own private Existence algorithm. Throughout all the turns, only systems which nature suited the Existence algorithm, have survived.

The spin of the electron around the proton is a physical rule at low resolution, that is derived from the Existence Algorithm, which is a physical rule at the fundamental resolution.

Today in observations we see systems of atoms and molecules, not because they are more correct systems among an enormous variety of possibilities of arranging the Existence in space and in depth, but only because these specific systems are adapted to survive in an algorithmic environment that prevails in our Private Universe.

In General Binary Physics, it may well be that the systems, which we are familiar with, would not survive even for a few turns in other universes with different Existence Algorithm, whereas other systems would be more common. There may even be universes, where the Existence Algorithm does not contain any system foundation, and therefore, no systems are formed at all. At the other extreme, there may also be universes possessing a strong system algorithm foundation, in which almost every system formed will survive.

History is written by the victors. In our case, we see only the victors in observations. When our mind is observing the Universe we sees only the patterns that survive, because their character suited the Existence Algorithm that prevails in our Universe.

"DNA" of Materials and Reproducing Materials

Each substance has its own "DNA".

> *"DNA" of matter: "The cyclical pattern of the Existence in depth that constitutes the properties that matter"*

The pattern of the Existence that constitutes matter in depth is, In analogy to the world of biology, is the "DNA" of the matter. Therefore, I will present again the pattern of points in depth, and now I will call them by their name, the pattern of matter in depth.

Simpler systems, such as atoms, have a simpler "DNA". More complex systems, such as molecules, have a more elaborated "DNA". Highly complex systems, such as human systems, have a very intricate "DNA".

The complexity of the system in depth does not change the basic principle: Every system has a unique pattern signature of the Existence and Placeholder arrangement in depth of which it is composed.

Once the pattern signature of the system in depth is damaged or decays, it is no longer the same system. The system ceases to survive.

For every system there is a unique pattern in depth.

If we project the unique pattern in depth of a particular material, we will receive it in space.

Once we are able to produce different patterns in depth, we can easily produce any substance that we want in space.

In fact, once we have an apparatus that will allow us to change the pattern of the Existence in depth, we can reproduce any material that we want, and even create new materials. With respect to new materials, since we cannot change the Existence Algorithm, we cannot ensure their stability. Nonetheless, I believe that a systematic scanning of patterns in depth will reveal other unique materials that are capable of surviving under our specific Existence Algorithm.

Such an apparatus is unequivocally possible from a physical standpoint, not just in terms of General Binary Physics, but also in terms of Private Binary Physics that governs our Universe and allows us to influence the pattern of the Existence in depth, and thus produce any material which we desire.

An apparatus for printing matter will be based on the Sixth Rule— that states the symmetry of space-depth.

Due to the symmetry of space-depth, it would be possible to create a particular pattern of obstructions in space, and thus influence the pattern of flow of Existence in depth. Any change in the depth pattern of Existence creates a new substance. On the basis of this principle, in the future we will be able to "print" any substance.

When Darwin published his book "On the Origin of Species by Means of Natural Selection, or Preservation of Favored Races in the Struggle for Life" in 1859 and made his revolutionary claim that animals develop in an evolutionary manner, no one imagined that not only is this true, but that within less than 150 years, scientists will decipher the human genome and be able to genetically engineer any living creature.

Similarly today, when we have gained the understanding that the properties of materials are determined by the pattern of a single particle on the time axis, it is difficult to imagine that in less than 150 years scientists will be able to reproduce any material on the basis of this assumption.

"Back to the Future" Hovercraft

Recognizing the "DNA" of materials in depth will help us build a hovercraft that can levitate over any material.

The principle that guides us here is the principle of magnetic repulsion. This principle is based, as we have seen above, on the formation of a pattern possessing an opposite slope in depth.

In order to form a pattern with an opposite slope in depth to that of the material on which we want to hover, and thus to create repulsion in space, we must first take the "DNA" (the depth cyclical pattern) of the system above which we want to hover. For example, the system of a water molecule.

After we calculate the depth cyclical pattern of the system of a water molecule, we can print matter possessing an opposite slope in its depth pattern, that will hover over the water, just as a magnet hovers over another magnet.

In a similar manner, we can form substances that will hover over any matter.

With even more futuristic thinking, it will be possible to create a device possessing the ability to regulate patterns, so as to match the opposite pattern of the matter over which it is hovering. It is possible that a "master" pattern will be discovered, which will be able to hover over all common types of matter in the universe.

Star Wars Light Saber

This is a sword that has the ability to cut everything.

To create a sword like this, we need that the system of "light" that is projected from it in depth will be such that when it encounters a depth pattern of a common system (for example, an electron), it will cause it to collapse.

The collapse of the electron's pattern will nullify its symmetry with the proton, and thus bring about the decay of the atom.

How can a depth pattern cause the pattern of the electron to collapse?

The pattern of Existence and Placeholder in depth is what determines the properties of matter—in this case, the electron.

The depth pattern of the electron is similar to this: 000001100000011 (in abstraction)

If we place opposite it a particle possessing a pattern exactly like this: 111110011111100 (in abstraction), then when they encounter each other, we will receive a situation of complete obstruction: 111111111111111 (in abstraction)

Damaging the depth pattern of the electron (destroying the "DNA" of the electron) will bring about decay of the entire system of the atom, and the result will be easily cutting through any substance.

Chapter 21
The Illusion of Speed and Time and the Source of Relativity

Since there is only one particle) I put aside the passive particle – placeholder), and every cell in the Universe has an identical response time (The Fifth Rule – Uniformity of the Queue), thus there is only one speed in the Universe.

From the physical aspect, this one "speed" that exists is called in Binary Physics "the speed of the cell response time".

> *"The speed of the response time of the cell":* "The Existence can move from one cell to another cell only at one constant rate, one cell in one turn. This rate is derived from the "response time of the cell". The response time of the cell in Private Binary Physics is one turn in the queue."

This being so, the word "speed", which in the language of human systems connotes a relative concept, is not the right one to use for a physical description of the phenomenon. When there is only one constant "speed" with no source for comparison, it is incorrect to use—from a physical standpoint—a word that implies a relative meaning.

Since there is only one physical speed, the speed of the response time of the cell, which is also the maximal speed, we should not be surprised to see one fixed and fundamental speed, when we look at the observations. Classical Physics calls this "the speed of light". This is also the maximal possible speed, as is known.

Therefore, the speed of light in Classical Physics is the speed of the response time of the cell in Binary Physics. And what is the definition of speed in general?

> *Speed: "A mental concept that reflects the ratio of change in space between one system and another"*

Speed is not a physical concept. It is a mental concept. It is a relative interpretation of consciousness at a low resolution of the movement of a system compared to another system. If so, why do we see systems moving at a "speed" slower than the reaction speed of the cell?

This is due to obstructions. But of course the objective is not obstruction in space, but obstruction in depth (what Classical Physics calls "the time axis").

An isolated Existence cannot slow down its motion, and therefore it cannot move at a slower speed than the cell's reaction speed. It has two states: moving, or obstructed. Therefore, as I mentioned before, the concept of speed is not relevant for an isolated Existence.

But unlike an isolated Existence, a system composed of many Existence definitely can "slow down" its motion.

A pattern of a point in depth at the fundamental resolution is represented as follows:

R(1)
d(4)0 **11011011** Initial State. The Existence algorithm is – "move to the right"
d(4)1 **10110110**

d(4)2 01101101
d(4)3 01011010
d(4)4 00110101

It is possible to see how the system marked in bold and being obstructed requires 4 turns to move one cell forward in terms of R(2).

That is, the speed of the motion of this system is 4 times slower than the speed of motion of the Existence without any obstruction.

It is understood that the wider the obstruction in depth is, the more the speed of the motion is slowed with relation to the speed of the reaction time of the cell.

There is an inverse relationship between the extent of the obstruction in depth and the speed of the system.

A system with no obstruction in depth at all—which is interpreted by us as an electromagnetic wave—moves as quickly as possible, at the speed of the response time of the cell (known in Classical Physics as the speed of light).

The more the system contains obstructions in depth, the more its motion is slowed down, directly proportionally to the extent of the obstruction.

The question of the relationship between matter and speed is the issue that has concerned me. Since there is only one speed, which is the speed of cell's reaction time, then necessarily what the mind interprets as different speeds must result from the different depth patterns of the systems. That is, a different speed in space = a different pattern in depth. If the pattern in depth of a system of a body that moves faster is identical to the depth pattern of a body that moves more slowly—then what is the explanation for the two speeds? Something significant in the depth pattern must change.

That 'something' is the level of the internal obstructions in depth of the system, which decreases as the speed increases.

The example of a traffic jam will assist us in understanding the point. The less congested the traffic jam is, the more the speed of the system of the column of cars increases, until it reaches the maximal speed, which is derived from the maximal speed of every car in the column.

When the system speed increases to the maximum speed, it essentially becomes free of obstructions altogether, and thus becomes pure change = pure energy = Electromagnetic wave

What is the relationship between the internal obstruction in depth the system and the speed of motion in space?

The Sixth Rule—the Symmetry of Space–Depth—requires that for every movement in space, there should also be a movement in depth.

Therefore, if the depth is obstructed and motion in depth is slower, the movement in space must necessarily be slower.

From this principle, a new concept has been born: The speed of time = vt.

> The speed of time (the rate of change in depth) = **vt** = the number of changes in depth relative to the number of turns.
>
> The maximum speed of time = **vtm** = one change in depth in one turn.

Since the speed in space is not uniform due to obstructions in depth, the speed of time is also nonuniform.

There is a direct correlation between the speed of time and the speed in space.

It can be said that the speed of time is also the relationship between mental time (relative time units of Classical Physics – the second, minute, hour *etc.*) and the objective depth units (the turns). The units of *t* are variable and are determined by the observer, Depending on his speed, and in any case relative. The units of the depth—the turns (*tu*)—are constant and absolute. Of course, the most basic (the fastest, if you like) unit of *t* is essentially one turn. The role of *vt* is essentially to bridge the gap between the basic unit of *t* as experienced by the mind and the turn. Therefore, it is possible to define *vt* as follows:

vt = *the number of turns in one mental unit of time*

For example, if we choose a popular mental unit of time, such as a "second", the speed of time would be equal to the number of turns occurred during one second. Of course, it is not a fixed number. As the system that the mind uses to measure a second (for example, a clock) moves at a greater speed, the greater the speed of time will be, and therefore there will be fewer turns in that second. In other words, the greater the speed of a system is, the fewer changes will occur in each mental unit of time (a second, for example), because fewer turns will transpire in each mental unit of time, and thus the mind will feel that time slows down. When the system accelerates to its maximum speed—the speed of the cell's reaction time (the speed of light) — mental time will stop. In every mental second, zero turns will transpire, and therefore there will be no changes.

For example:

In a system with slow Spatial speed, 1 psychological second = 100 turns
In a system with fast Spatial speed, 1 psychological second = 50 turns

Thus, in a fast system, fewer changes can occur "in a second", and thus "aging" is slower.

The speed of time is an absolute physical concept, because it is derived from the two absolute physical concepts: change and the turn. This is distinct from conventional time, which is a relative mental concept that will be defined shortly.

Since vt is also the rate of change, it is more correct to call this concept simply **the rate of change in depth**. However, because we are still so used to concepts of Classical Physics, I chose currently to remain with the less precise name, "the speed of time".

The faster the speed of time in a given number of turns is, the greater the number of changes in depth will be.

The slower the speed of time in a given number of turns is, the less the number of changes in depth will be.

When there is no obstruction in depth, the flow in depth is performed at the speed of the cell's response time, which enables, in accordance with the Sixth Rule, motion at the maximal speed in space. This flow of the Existence is what we call light or electromagnetic radiation. In such a state, with no obstruction, a maximum of changes in space are made possible in a single turn. This is the state in which the speed of time is the maximum: $vt = c$, the latter being the speed of response time of the cell.

When there is an obstruction in depth, fewer changes are possible in space in each turn. The speed of the system in space decreases. In such a state the speed of time vt is lower.

In a state of the complete obstruction in depth, no changes are possible in space in the queue. The system is "frozen". The speed of time in this state $vt = 0$.

The speed of the system in space increases = the speed of the flow in depth increases (by the Sixth Rule) = Change (energy) in every turn increases = the internal obstruction in the depth of the system (mass) gets smaller and/or the relationship between the obstructions and the placeholders within the depth pattern of the system is smaller = the speed of time is greater.

I will note that it is more correct to write the opposite: the speed of flow in depth increases = the speed of the system in space increases. This is because the speed in depth is what determines the speed in space, not *vice versa*.

The speed of the system in space is maximal (speed of light) = speed of flow in depth is maximal (that is, the speed of the cell's response time) = change (energy) in each turn is maximal = there is no obstruction in depth (no mass) = the speed of time is maximum.

The speed of the system in space is low = speed of flow in depth decreases = change (energy) in each turn is small = the quantity of internal obstructions in the depth pattern of the system (the mass) increases and/or the relationship between the obstructions and placeholder in the system's depth pattern increase = the speed of time is lower.

The system is frozen in space = the speed of flow in depth is zero = change (energy) in every turn is zero = complete obstruction in depth (mass is maximized) = speed of time is zero = time stands still.

What is Time?

> *Time: "A mental concept that reflects the ratio between the rate of change in depth of one system and another."*

Without change, time stands still, but the queue never stops.

From a physical standpoint, there is no such thing as "time". There is the queue.

The queue is uniform and absolute; this is due to the Fifth Rule.

The concept we call "time" is a mental concept. Time is the way in which our mind interprets the ratio of change in depth between two different systems.

Therefore, we can speak of the speed of time, but we cannot speak of the speed of the queue.

In Classical Physics, when we talk about time, we actually examine how many changes have happened in one system, as opposed to how many changes occurred in a second system. For example, how many changes occurred in the system of my wristwatch as compared to how many changes occurred in the system of a car. The Classical Physics definition of time is relative.

As the speed of a system increases, the speed of time and the rate of change increases, whereas the ratio between the maximum rate of change (that is based on the response time of the cell) and the rate of change of the system decreases. In such a state, an interesting effect is created.

The Theory of Relativity in the View of Binary Physics

Example: A human system and a clock system are moving in space at the maximum possible speed = the speed of light = the speed of the response time of the cell.

Note: As I mentioned earlier, systems that possess internal obstructions in their depth pattern (mass) are not able to move at the speed of the response time of the cell—because movement at this speed requires a depth pattern without obstructions - that is, pure change (pure energy). Therefore, this is a theoretical example, that is intended for purposes of illustration only.

Now, the human system looks at his watch. He will see that the system of his watch has stopped. Of course we know that once the speed in space is maximum, the speed of time is maximal, and therefore the changes in the system of his clock are maximal in relation to the queue. Then how is it possible that the human system' mind reports an observation that the clock has stopped?

The source of this illusion is the relative nature of time.

Time is defined as the ratio of the rate of change in depth between one system to another. When the speed of time of the system of a clock is maximal, and the speed of time of a human system is maximal, the rate of change in depth of the two systems is identical. The difference in rate of change in depth between the two systems is zero, and there is no way of measuring the time between them.

When the human system moves at a low speed, a difference is made possible in the speed of time between it and the system of the clock, which parts are moving at high speed (such as the quartz crystal). When the human system enters with the system of his watch into a spaceship and they start to accelerate, the difference between the speed of time of his system and the speed of time of the system of his watch will decrease, as both of them will accelerate more and more towards the maximal speed. The mind of the human system will report to him that the speed of the clock's hands decreases with respect to him, until—when the two systems will reach the speed of the response time of the cell—the mind of the human system will report an observation of an absence of change in the system of the clock. When the two systems, the human system and the clock system, will move at the maximal speed, the speed of time of both will be maximal and will not allow for measuring the time of the one system in relation to the other. Time will seem to stop moving, despite the fact that its speed will be maximal.

When the speed of time of the two systems is maximum (there are no internal obstructions in depth), it is not possible to measure time between them, and from the viewpoint of one system with respect to the other time stands still.

When the speed of time of the two systems is zero (there is a total obstruction in depth), it is not possible to measure time between them, and from the viewpoint of one system with respect to the other time stands still.

Imagine two train tracks.

The one on the right will serve as a clock for the one on the left.

When the left train is stationary, a passenger in this train will count in one hour 100 trains that pass by on the right track.

As the left train increases its speed, the passenger will count fewer and fewer trains that pass by on the right track. When the left train reaches half of its maximum speed, the passenger will count 50 trains on the right side. When the left train reaches its maximum speed, the passenger will count 0 trains on the right side. That is, from his viewpoint, time has "stopped".

The trains simulate the movement of particles in depth. The left train represents an accelerating system, and the right train represents the system of the clock. In this example, we see that there are two situations in which it is not possible to measure time:

The first: the two trains are moving at maximal speed.

The second: both trains are stationary.

The system of time measurement—the system of the clock—is like any other system. Therefore, when its spatial speed increases = its speed in depth increases = its speed of time increases = there are fewer turns in each psychological second = there are fewer changes in every psychological second = psychological/mental time slows down.

Suppose that on the Earth's surface, the speed of time of the system of my watch is 10 times bigger than the speed of time of my own system. In such a state, I will be able to measure the change in my system versus the change in the system of my watch.

As soon as my watch and I will go into a spaceship and begin to accelerate, and the speed of time of our two systems will increase, the gap between the speed of time of the two systems will be reduced, and with it, the ability to measure time.

If on the Earth's surface the ratio is 1:10 in favor of the clock when the spaceship begins to accelerate, the ratio will decrease to 1:7 (I will age slower) and then 1:3 (I will age even more slower), and in the end when the two systems (those of myself and of the watch) will reach the speed of the cell's reaction time, the ratio will go down to 1:1 (I will not age; time from my perspective will stop). When my system and the system of my watch will move at the maximal speed of time, I will not be able to see any change on my watch.

At the maximum speed of time, there is no obstruction at all. The internal obstruction (the mass) and the external obstruction (the force) is the causes of changes in systems. In other words, without obstructions there will be no changes in my system. Without change I will not age at all. As long as my system does not encounter an obstruction and not slows down its speed, it will remain young forever…

You could say that when I increase my speed, I encounter less and less obstructions so it's like I go into a type of a time freezer. My change with respect to myself is frozen. There is no obstruction that would change my system.

Of course, the queue continues and therefore systems at lower speeds (such as the system of my twin brother who remains on the Earth) continue to change. Therefore, many changes will be created in the system of my twin brother on the Earth (changes that we perceive as negative – the changes of aging) after 100 turns, while my system will undergo no change. Therefore, even though in absolute physical terms, the same number of turns—100— will transpired for my system and the system of my twin brother, many more changes will be created in my brother's system, and therefore the illusion is created that my system remains young (without change), while my twin brother system has aged (many changes).

Since time is a relative mental concept that compares changes in different systems, the most strong feeling of time happens "in the middle"– between zero speed of time and the maximum speed of time.

Zero Speed of Time = Zero Spatial Speed = total obstruction in depth = the system remains unchanged = time stands still.

The "medium" speed of time (under the term "medium" I refer to the middle of the scale between zero and maximal speed of time that based on the cell response time) = "medium" spatial speed = partial obstruction in depth = changes occur to the system = the relationship between the changes in the system and the lack of change is maximal = the mind interprets it as a feeling of maximal time = maximal psychological time.

Maximum speed of time (at the rate of the cell's response time) = maximum spatial speed = lack of obstruction in depth = the pattern of the system remains unchanged due to a lack of obstruction = psychological time stands still.

If an astronaut is launched into space at the maximum speed (the speed of light) for 1000 turns, we will feel many changes on the Earth, due to obstructions that our patterns encounter. These are the same obstructions that originally cause us on the Earth to be at a lower speed of time. The same 1000 turns will transpire even for the astronaut that travels at the maximum speed. But since he will be in a state of lacking obstructions (which is the state that is required originally in order to travel at the maximal speed), his pattern will be totally preserved, and after 1000 turns he will not feel that he had experienced any change. This is a theoretical example, for purposes of illustration, since there is no possibility of accelerating a system possessing internal obstructions (mass), such as an astronaut, to the speed of the reaction time of the cell (the speed of light) without the obstructions that comprise the depth pattern of the astronaut breaking up.

> *As the speed of time of a system increases, the quantity of obstructions that it contains in it's depth pattern decreases.*

Two processes start to occur in parallel:

First, the amount of change of the system increases (meaning the quantity of Existence that is flowing in depth in relation to the number of turns).

Secondly, the system's pattern changes less and less, because it contains fewer and fewer obstructions.

During this process, the ratio between the quantity of changes in the system in Compared to the possible changes in other systems is maximized—the perception of time is a maximum.

You can say that Einstein's Theory of Relativity stems from the principles of flow in depth.

Acceleration

The very definition of a uniform speed negates the meaning of the concept of acceleration. After all, from the outset, how can you accelerate when you are moving at maximum speed (or minimal speed, or at any arbitrary speed—as I mentioned, the instant that there is a single speed, there is no meaning in the concept of speed)?

Classical Physics defines acceleration as follows: "the rate of change of speed of a moving body"

Since, from a physical standpoint, there is only one uniform speed, and there is no acceleration (or deceleration), thus, the concept "acceleration" is not a physical concept, but rather a mental concept. This is the interpretation of our mind, at low resolution, of the change in the rate of change in space.

It can be said that when there is change in the speed of time, the mind interprets this as acceleration or deceleration.

It is possible to speak of acceleration—precisely as about speed—only when you give an interpretation at low resolution of motion of the system.

At the highest resolution—the fundamental resolution—there is no "acceleration". Every Existence particle can move or be obstructed. Once it moves, it moves at the speed of the response time of the cell, i.e. one movement in one turn.

> *Acceleration: "A mental concept that reflects, at low resolution, the increase in the rate of change in depth, and as a derivative of this, the possibility of increasing the speed in space."*

And if we will use the concept of acceleration, then the concept of deceleration is more relevant. The fundamental force—the obstruction—is not an accelerating force, but rather a "slowing" force. An obstruction "slows" the speed of the Existence flow in the Universe.

How can I use the word "slows", when I have just determined that there is no meaning to the concept of speed?

After all, just as it is impossible to accelerate the speed of the motion of the Existence in the Universe, it is equally impossible to slow it down. The Fifth Rule determines that the response time of all cells is the same.

Slowing down due to obstruction does not affect the speed of an isolated "Existence". It can of course block it from moving, but it cannot slow it down.

The process of slowing down reflects only in the observer's mind when it views a large system of the "Existence" at low resolution, and feels that the entire system moves at a slower speed than the speed that is derived from the response time of the cell.

> *The smaller the obstruction in depth = the speed of time increases = the spatial change increases = spatial speed increases = there is acceleration.*
>
> *There are no obstructions in depth = the speed of time is maximum = the spatial change is maximal = the spatial speed is maximal (the speed according to the reaction time of the cell = the speed of light) = there is no possibility of further acceleration.*
>
> *The more the obstruction in depth increases = the speed of time is less = the spatial change is smaller = the spatial speed is less = there is deceleration.*
>
> *Maximal obstruction in depth = the speed of time = 0 = There is no spatial change = spatial speed = 0 = there is no possibility of deceleration.*

To understand how the obstruction creates deceleration, we can imagine looking from a helicopter on a traffic jam at low resolution. We will see a long and colorful snake (rather than separate cars) traveling at a speed that is much less than the maximum speed at which each car can move separately.

This example is just for purposes of illustration. There is a great difference between cars and the Existence. The Existence, as a physical concept, can move only at one speed, and the cars, as large systems at low resolution, are capable of accelerating from a low speed to a higher speed. Therefore, if you want to be precise in the example, you have to imagine a car that can either stop or move at 100 km/h without any option to move at a different speed. Such a binary car...

Newton's First Law

Newton's first law states that a body will remain in its state, as long as no external force is applied on it. This is the rule of inertia.

As we have seen, the physical truth is the opposite. According to the Seventh and Eighth Rules of Private Binary Physics, every Existence or system of the Existence is found to tend to move to the maximum possible speed, and only when a force is exerted upon it (an external obstruction) will it slow down...

Newton's Second Law

Newton's second law is $F = ma$. Force = Mass * Acceleration.

Binary Physics agrees with this assertion, despite the formulation of the rule in Binary Physics is opposite. In Binary Physics the value of the force is a mirror of the value of obstruction.

Acceleration is formed in one of two instances:

1. Reducing the internal obstructions in the system and converting it to change = converting mass to energy. It follows from this that when the internal obstruction (mass) is small, the acceleration increases. There is an inverse relationship between the internal obstructions and the acceleration.

2. Reducing the external obstructions to the system = The force increases (the value for force in Classical Physics is in an inverse relationship to the value of the scope of the external obstructions. As the value of the scope of the external obstruction is greater, so the value of the force is less). It follows from this that when the external obstruction is small, the acceleration increases. There is an inverse relationship between external obstructions and the acceleration.

Therefore, the formula for acceleration in Binary Physics will be:

Acceleration = The value of reduction of external obstructions * The value of reduction of internal obstructions.

In Binary Physics, Force is the external obstruction of the system, and in terms of Classical Physics, when there are fewer obstructions, there is more "Force". Therefore, as opposed to Classical Physics, in which, as the value for the force increases, the acceleration increases, in Binary Physics, as the value of the obstructions decreases, the acceleration increases.

An example for the sake of illustration: Balls that flow through a pipe. As the diameter of the pipe is larger (that is, the external obstructions are fewer), and as the diameter of the balls is smaller (that is, the internal obstructions are fewer), the flow will thus be more rapid.

From Classical Physics, it is known that as the object accelerates, thus its resistance to the acceleration increases; that is, more force is required in order to accelerate it. In other words, its mass gets larger.

According to Binary Physics, as the system accelerates, the internal obstructions are removed, and thus the mass decreases. A system that moves at the maximal speed necessarily possesses a depth pattern lacking internal obstructions. Apparently, there is a contradiction here.

But, this is not a paradox. A system can accelerate in one of two ways. The first is if the internal obstructions in its depth pattern are opened, the Existence flows in depth more quickly, and as a derivative of the Sixth Rule, spatial change also flows and becomes quicker.

The second, and it is this that solves the paradox: in systems at low resolution, it is possible that the flow is faster in depth even without damaging the pattern of obstructions, but rather the quantity of Placeholder between each obstruction is reduced.

The depth pattern of System A: 111000001100000111000011 (Slower)
The depth pattern of System B: 1110001100011100011 (Faster)

It is possible to see that the depth pattern of System B is the same pattern as system A (in terms of the length of the sequences of obstructions), but the ratio between the obstructions and the Placeholders between them is lower. Several conclusions follow as a result:

1. The flow of System B in depth will be quicker, and therefore, according to the Sixth Rule, it will allow for a greater speed in space.

2. System B is more energetic.

3. Despite the fact that the length of the sequences of obstructions between System A and System B are identical, System B contains more obstructions relative to its size in depth than

System A, and therefore it is possible to say that its mass is greater.

When a system possessing mass accelerates, in order that it should remain similar to the original system, its depth pattern of obstructions must remain, and the change in its pattern will focus on reducing the percentage of Placeholder between obstructions. It is understood that this rule puts a boundary to the maximal speed to which a system possessing mass can accelerate. Beyond a certain critical threshold, after the system has "used up" the reduction of Placeholder and the internal obstructions collide with each other, in order to allow the system to continue to accelerate, its obstructions in depth are required to break up, in accordance with the Tenth Rule (the system loses its mass), and in fact it is possible to say that from this instant the original system as a system possessing obstructions (mass) ceases to exist, and it converts into pure energy (for example, electromagnetic radiation). We see this phenomenon for example in atomic explosions, when part of the mass of the original system changes over the course of the explosion to the energy of electromagnetic radiation. The system of electromagnetic radiation is different from the original material that served to provide the explosion.

The more that it is necessary to reduce the percentage of Placeholder that is internal to a particular system, thus it is required to perform a more significant removal of the obstructions that are external to the system (that is, greater force, reminding us that reducing the external obstruction increases the "force").

Newton's Third Law

This rule states that when a body exerts any force on another body, the second body will exert a force equal in strength on the first body, but opposite in direction.

In Private Binary Physics as well, when the Existence or a system obstructs (exerts a force) on a second Existence or system, the second Existence or system obstructs (exerts a force) at the same time the first system.

Speed and Resolution

So how, ultimately, does the mind feel (observe) different speeds? In fact, the mind does not feel different speeds, but rather it feels different resolutions.

When a system moves quicker, so do we perceive it at a higher resolution.

When a system moves slower, so do we perceive it at a lower resolution.

When our consciousness observes a beam of light, moving at the maximum speed, it perceives reality at the maximum resolution. It interprets the maximum resolution as light.

When our consciousness observes the system of a tortoise, moving at a slow speed, it sees the reality at low resolution. It interprets low resolution in depth as different types of matter.

When our consciousness observes a system, that is found to be totally obstructed (at the temperature of absolute zero), it sees reality at the lowest resolution of which it is capable of seeing.

Go and learn: Speed is resolution. The two concepts are one and the same.

> *Higher speed = higher resolution*

Speed and Spatial Size

> *Contraction of the system: "Decrease in the percentage of Placeholder in the system"*

We saw that for a system to maintain itself when increasing resolution, it must contract. We also saw that the increase in speed means an increase in resolution. Therefore, a system whose speed increases – contracts.

But how is this contraction created?

The Sixth Rule states that there is simultaneous movement of the Existence in space and in depth. In order that the speed in space will increase, the speed of time (the rate of change in depth) must increase. In order that the rate of change in depth will increase, and the depth pattern of the material (which determines its properties) will be preserved, the percentage of Placeholder must decrease. Practically speaking, the system contracts in depth. And in a circle, the Sixth Rule obligates that the contraction in depth will cause contraction in space as well.

The pattern of a point in depth before the speed increased:
1111000011110001111
The pattern of a point in depth after the speed increased:
1111001111001111

We see that after the increase in speed, the depth pattern contracted from 16 cells to 12 cells. The pattern itself is preserved in terms of 3 sequences of 4 Existences, but the percentage of Placeholder has been reduced. According to the Sixth Rule, the contraction of a pattern in depth will necessarily cause a contraction of the pattern in space, with an increase in speed. In reality, these sequences that I mentioned earlier represent durable sub systems within the system.

> *Increase in speed = increase in the speed of time = increase in the rate of change in depth = contraction of the system in depth = contraction of the system in space.*

When something moves at a faster speed, its **rate of change** in depth increases, and in symmetry with it, the **rate of change** in space also increase. That is, its density increases. Instead of expressing of its information by means of 100 cells, I can now express all the information by means of 50 cells. The spaces between the cells of the Existence decrease.

It is not possible to increase the spatial speed without creating a contraction in space. This is because, in order to increase the spatial speed of the system, I need a higher **rate of change** in depth (in order that the system will move faster, its speed of time must be increased). A higher **rate of change** in depth requires, in accordance with the Sixth Rule, a higher **rate of change** in space = more dense space.

Of course, there is a limit to the density, and therefore there is also a limit to the speed of system with internal obstruction (mass).

Beyond the level of a particular density in depth, the sequences of "Existence" have collided with each other and have broken up, in accordance with the Tenth Rule, and then the system essentially will not be the same system. There is a limit to how much it is possible to contract a system possessing internal obstructions without dismantling or obstructing it.

To illustrate this, let us take a computer screen that displays a face by means of, for example the Microsoft program "Paint". If we look at the system of the face at higher resolution by increasing the zoom—the size of the display on the screen—at a certain point, we will not see the face on the screen, but rather a collection of pixels that display only a part of the face. The moment that we do not see the entire face, we will imagine the pixels that we do see on the screen as another system.

So that the system of a face should remain, even while resolution increases, the image must contract together with the increase in resolution. As long as it is possible to contract the spaces between one point and another in the pattern of the face, without changing the pattern itself—I am able to increase the resolution and still see exactly the same face, without losing any information.

Chapter 22
The Secret of Gravity

Classical Physics defines gravity as follows: "An attractive force that acts between two bodies in accordance with the product of their masses and their distance from each other."

Einstein, in his General Theory of Relativity, stated that the phenomenon of gravity is due to the warping of space-time. This is because all the mass causes a warping of space-time around it proportionately to its size. Each body aims to move from point to point in the shortest path. Due to the warping of space-time, the body's trajectory will appear warped to the observer. This curvature creates in the observer's mind an illusion of an imaginary force, which causes the body to move in that curved path. The force of gravity is essentially that imaginary force.

A Star

Because gravity is a force, it must be derived from the fundamental force – the obstruction.

And so it is.

Our sense of the curvature of space-time is due to the difference in the power of obstruction in depth in different regions of space.

> *Star:* "*An opening in an obstruction in depth, through which the Existence flows.*"

The pattern of the star, such as the Sun, which we see in space, is a kind of negative of its complete pattern in depth.

I will explain by way of example. Since we are blind to depth, and it is difficult to talk about four dimensions in relation to three dimensions, we will take an abstract example of two dimensions as compared to three dimensions.

Imagine a sink with a drain in the center at its bottom. All the water will flow into the center of the sink. Suppose that a two-dimensional mind exists in a drop of water in the sink. It senses forward/back/left/right. This mind is blind to downward/upward movement.

When the mind reaches the region of the drain, it will report of a "force" that pulls it to its center in a circular motion. The mind will not see that it also descends, while moving toward the center of the sink, to its depth as well.

As the drop approaches the center of the sink, it will report of a strengthening of the force and an increase in the speed of its motion toward the center of the sink.

In the center of the sink, the "force" that it will feel will be at its peak.

The structure of the sink is analogous to the pattern of the star in depth. The part of the pattern that we do not see. The shape of the water around the opening of the drain is like the star that we see in space.

The star is a huge obstruction in depth with an opening in its center.

The area in space, where we see the planet, is actually the area of the opening. Therefore, we see a huge flow there (which we interpret as the star in space) and a huge change (tremendous energy).

A kind of reverse mode is created. Precisely in an area with no blockage in depth, we see the maximum obstruction in space. Just as in the case of the sink, precisely when there is the opening of the drain, we see the peak flow of water.

All Existence is trying to move to the next level of depth through the area with no obstruction in depth. Thus, where will we see the most energetic motion of the Existence? The greatest change, or in the language of Classical Physics, the most energetic region? Exactly at the entrance of the obstruction in depth. Therefore, at the center of the star, which is the opening in depth, we will see the greatest amount of change. In terms of Classical Physics, we find the most energetic state—the greatest heat—at the center of the star.

When the Existence, or a system of the Existence, arrives at the region of a star—the area in which there is a large obstruction in depth—it starts to revolve in the direction of the opening of the obstruction, i.e. towards the center of the star. This is the force that we interpret as gravity.

Of course the force of gravity acts on huge systems such as stars, but also on a system of any size.

From the Birth of a Star to its Death in a Supernova

At the beginning, there is a huge, concave obstructing system in depth (like a sink) with an opening in the obstruction. The Existence particles in space that reach the area of the huge obstruction are affected by it and start to revolve around it in accordance with the Tenth Rule, while they are slowly gliding in depth towards its center in the direction of the opening (of course, our mind, which is blind to the depth, sees only the motion in the direction of the center, not gliding in depth).

This process causes more and more Existence to accumulate and move toward the opening, until eventually a sort of "traffic jam" of the Existence is created and is swirling in a vortex around the opening. The star is this vast quantity of the Existence in space, which spins around the opening and waits for its turn to pass through it.

The center of the star is the area of the opening in depth, and the entire star is the "traffic jam" of a huge quantity of the Existence that is advancing toward the opening.

Therefore, we see the greatest activity—the greatest change (the most heat/ energy) — in the region of the center of the star, the center of the opening.

The planets that orbit around the star are (relatively) durable systems of Existence that flow along the edges of the star's depth pattern.

Since the planets are durable systems (systems whose system algorithm does not allow an external existence to enter into them and destroy their pattern), they do not mix with the star's system, but simply "glide" along its depth structure. In space, ignore the dimension of depth, we see the spinning of a planet around a star and explain this as "gravitational force".

As the queue progresses, the huge quantity of the Existence flowing toward the opening of the obstruction simply starts to "clog" it.

The opening in the obstruction starts to close.

As the opening in the obstruction starts to close, the change around the opening starts to decrease. Classical Physics interprets the decrease in change as the star beginning to use up its energy. True, change is created as a result of the presence of an opening in the obstruction, and therefore at the onset of the star evolution, change is at its peak (it is full of energy). At the end of the star's life, when the opening starts to "seal" more and more, change decreases (and our mind interprets this as if its energy is used up).

Eventually the opening is completely obstructed, or its obstruction reaches such a level that it adversely affects revolving of the Existence around it. At this instant, all the Existence particles that stop "being drawn" to the opening in depth, start to scatter in all directions, and we will see a supernova.

Imagine a drainage hole that suddenly becomes clogged, and the water in the garden stops its orderly flow in its direction, and starts to scatter in all directions and floods the garden… A supernova of water in the garden.

I must clarify something about the example of the sink and the water. Water is a substance that normally we imagine as uniform. The Existence that flows along the obstruction towards the opening is not uniform. At the region of the opening it is the densest—very congested. There is almost no Placeholder. And as we move away to the area of the edges of the obstruction, it becomes much less congested, until we see huge spaces between one cluster of the Existence and another—while we call these distantly separated clusters, "planets".

Gravity in Binary Physics

> *Gravity: "The influence of the obstruction in depth on the direction of motion of the Existence in space"*

As we have seen earlier, the obstruction is mass as well as force. When do we interpret the obstruction as mass, and when do we interpret it to be a force?

> *When the obstruction is internal to the system—it is interpreted by us as mass.*

> *When the obstruction is external to the system—it is interpreted by us as a force.*

For example, the depth pattern of the system of a star, which forms obstructions that influence the movement of a planet's system, is interpreted as gravitational force. This is an obstruction that is external to the system.

Gravity is actually the pattern in depth of any system in space. The pattern in depth of the system is significantly larger than its pattern in space.

Within the huge obstruction in depth, such as the obstruction of a star, there are smaller obstructions, such as the obstructions of planets, that cause systems of Existence, such as the moon, to spin around them.

It is important to understand that ultimately every system with mass have a pattern of internal obstruction in depth: from human systems, through the system of an apple, to the system of the atom.

Gravity and the Tenth Rule

The Tenth Rule states that when the Existence encounters an obstruction in depth, it will bypass it in a circular motion in space to the next level of depth.

The Tenth Rule is the primeval source of the rotational motion, which is so common in the Universe, but is also the source of gravity.

When the Tenth Rule operates on a system at low resolution, it causes a spinning of one system around another.

Mass is an internal obstruction in depth of the system.

When a system with little mass approaches the area of a system with a large mass in depth, it starts to spin around it in depth, in order to bypass it.

The human mind is blind to depth, and therefore we must not become confused. The concept "bypass it" is intended in depth and not in space, which we see in space as the spinning of one system around another system.

You have to remember that the system that we see in space is only a small part of the whole system, most of which is found in depth. It is similar to an iceberg, of which the observer floating on the ocean's surface sees only the tip.

When we look at the Sun, we see only a small part of its system. Most of its system is found in a lot of spatial cubes in depth and is distributed over a much greater area than what we see.

In fact, the Sun that we see is part of the structure of the Sun in depth, in which there is no obstruction, or in which the obstruction is very minimal. Therefore, we see all Existence in our space drain out to an area, where our mind sees it as "Sun".

According to the Seventh Rule—trivial motion in depth—the Existence always tries to advance in depth.

When there is a large mass in depth (= a large obstruction), the Existence in space will try to bypass the obstruction, in accordance with the Tenth Rule.

Assuming that the obstruction in depth is not uniform, and there are "holes" (regions without an obstruction or with a minimal obstruction), all Existence in space will drain out to those regions, in an attempt to advance in depth.

Paradoxically, precisely in the areas where we see the greatest mass—such as in the Sun—there we have the least obstruction in depth.

Think again about the porcelain sink filled with viscous material, which can change its density. The material is much less dense than the porcelain, and therefore cannot pass through it, but only through a small drain opening located in the middle of the sink. We will see the greatest flow of the material at the drain opening. There the material will also be the densest.

The Flat Ones (that can only see the material in two dimensions and cannot see the porcelain sink that is a three-dimensional structure) will state that the greatest and denser mass of the material can be found in the center of the space (where the drain opening is, which they cannot see).

Understandably, the interpretation of the Flat Ones is erroneous. The most significant mass is actually located in the structure of the sink itself, not including its center. The central mass, the central obstruction, is the porcelain. Precisely where they see the concentration of mass—in the area of the drain opening—there is no obstruction.

The moment that more viscous material reaches the area of the sink from the bottom of the sink—the structure of the sink will influence it, and it will start to spin around the center of the sink (the place of the drain opening) until it reaches it and goes through it.

The Flat Ones will interpret the sink area as a region in which a force acts that causes mass—additional viscous material that comes from the bottom of the sink—to start to be drawn in and spin around the central mass that is located in the center of space (the area of the drain opening).

The Flat Ones, as two-dimensional beings, will not see the true reason for the spinning of one mass around the larger mass. They can only infer it logically.

We, human systems, are the Flat Ones. We are blind to depth, and therefore do not see all the mass of the Sun that is distributed in depth over the entire area of the Solar System. The region in the center of the Solar System in which we see the Sun—is essentially the region in which there is a "hole" in the Sun structure in depth. Therefore, all of the Existence in the space of the Solar System, that tries to bypass the huge obstruction of the Sun in rotational motion, will drain out in the direction of the Sun (the "hole" in depth).

The masses that we see in space are drain openings of depth.

The masses that we see in space are a negative of the true distribution of mass in depth.

Where we see mass, in fact, there is a hole in the obstruction in depth.

Since the Initial State in Private Binary Physics was the condition of high density in both space and depth at the origin of the Universe, an example of rotational motion of the Existence around regions with less obstruction was created.

Black Holes

The black holes that we see in space in the center of almost every galaxy are essentially openings in huge obstructions that extend in depth over the entire space of the galaxies. Our mind does not see all of the structure in depth of the galaxy, but only the movement of the solar systems in a circular motion toward the center of the galaxy—to the opening in the obstruction at the center of the galaxy.

You could say that, in a sense, a black hole is just the opposite of an obstruction in depth; it is a hole in depth. It is as it is called… instinctively human systems search for the hole in space… but the hole is in depth…

Since black holes are openings in the obstructions that are so enormous in depth, the flow of the Existence in depth in them is enormous. The speed of time in them is the highest.

In a region saturated with the force of gravity, i.e. a "hole" surrounded by a large obstruction in depth, the Existence flows at the greatest speed, and the speed of time is the greatest there.

If we imagine a situation where the mass in depth is sealed without any "holes", all motion in space would be frozen. The Sixth Rule obligates that all motion in space is accompanied by simultaneous motion in depth.

Two parameters affect the force of gravity.

First, the width of the obstruction in depth. The wider the obstruction is, the more the force of gravity is distributed over a wider area.

Second, the slope of the obstruction in depth. Going back to the example of the sink, the more that the sink is sloped/steep toward the opening, the more quickly Existence will flow in it (or rather, it will slow down less the speed of the response time of the cell). We will see this as a faster speed of time, and as a greater gap in space between the speed of time from one place to the other.

A black hole is also characterized by a very wide obstruction in depth, and a steep slope of obstruction in depth.

It is possible to imagine a black hole as a very small opening in a water pipe, towards which a large quantity of water flows. The flow that leaves the opening will be very strong. The mind will interpret this as immense gravity. As we widen the opening, we decrease the slope around it and decrease the water pressure, so that the flow that goes out through it will be weaker. The mind will interpret this as low gravity.

Again, I find it fitting to mention that our mind, being blind to depth, essentially does not see the pattern of the black hole in depth, but only its expression in space.

The Rotation of the Moon

The distribution of the Existence in depth causes many secondary flows. There is the central flow that results from the central hole in the mass of the galaxy in depth (the black hole). But within the central flow, there are infinite secondary flows of mass that try to bypass another mass in depth. The mass of the Earth that tries to bypass the mass of the Sun, and the mass of the Moon that tries to bypass the mass of the Earth.

In fact, the two masses that we see in space are two flows in depth that are caused by holes in depth (regions of a lack of obstruction in depth) that spin around each other.

At the lowest resolution, we can compare the entire Universe to one huge whirlpool of bypassing in depth, and as we increase the resolution, we will reveal in this great whirlpool a huge quantity of secondary whirlpools—planets that revolve around stars, moons that revolve around planets, until the electron that spins around a proton.

Newton and the Apple

When we throw a regular apple and an identical apple made from iron from a tower, why will they both reach the ground at the same time?

This is because the two systems glide along the edges of the obstruction of the Earth, and their motion is influenced by the pattern as well as the slope of the obstruction in depth, and not by their own structure. Just like two different children that slide down a water slide will reach the bottom at the same time. This is assuming of course that both are influenced only by the speed of flow of the water on the slide and the structure of their bodies does not create friction with the slide itself.

Wormhole

A wormhole is an area that possesses a higher resolution, and thus the speed of time in it is higher, and therefore it is possible, by means of it, to move more quickly in space.

Again, I mention that, contrary to the common sense, the Universe is not constructed with a uniform resolution, but rather composed of areas with varying resolutions. Our brain interprets all this as a uniform image.

In our brain's interpretation, the high resolution is interpreted as a higher speed.

Space itself is constructed of varying resolutions. And therefore in different regions it is possible to move at different speeds.

Chapter 23
Time Travel

In terms of Binary Physics, time travel means the motion of the Existence in the fourth dimension—the dimension of depth. From a physical aspect, it is possible to say not only that time travel exists, but that it is axiomatic and follows directly from the Fifth Axiom.

At the fundamental resolution, every Existence that passes from one level of depth to another performs what the mind calls "time travel".

There is one situation in which the Existence cannot perform its travel in time—that is the situation of the total obstruction. In this situation, the Existence is frozen in time, not only mentally, but physically.

In Private Binary Physics, the sixth, Seventh and Tenth rules require the Existence to move in depth in a particular direction. Therefore the depth before the present depth becomes "the past", and the depth following the present depth becomes "the future".

This situation is not obligatory in General Binary Physics.

You can certainly imagine a universe in which the Existence can move freely in depth in any direction.

In such a universe, there would be no definitions of the past, present, or future. We would be able to see system of the Existence from a more progressive depth approach us. Your great-great grandson would pop into your house to ask for a glass of milk...

In our private Universe, observations indicate that there is a uniform direction to the progress of the Existence in depth, and there is the Sixth Rule (Trivial Motion in Depth), and therefore we do not see systems "from the future" coming to visit us...

A Journey to the Past

How is this possible?

> *Physical past: "Physical spatial cubes that precede the reference cube"*

> *Mental past: "The mind's memory about the character of the spatial cubes preceding the reference cube"*

The mental past is not located in "the past". It is found in the present. Memory is an arrangement of the Existence at low resolution that is located in the present spatial cubes, in which there are memory storage devices. Depending on the reliability of the memory's standard, the mental past is fixed, as opposed to the physical past that is subject to unceasing change.

In Private Binary Physics, it is not possible to move toward the physical past. This is impossible, in accordance with the Sixth, the Seventh and the tenth rules, which explicitly determine the direction of motion of the Existence in depth.

In General Binary Physics, there is no obstacle to the passage of the Existence to the previous spatial cube. For example, in a universe with the Sixth Rule formulated in such a way that the movement of the Existence in space, necessitates its passage to the next or the previous depth.

In General Binary Physics, in which the Fifth Rule is not obligatory, regions of space are possible where by definition the motion of the Existence in depth will be faster or slower—something that will allow for easy transition between spatial cubes. More precisely, in a situation where the Fifth Rule is not required, it is not correct to speak of spatial cubes as an independent concept, but rather that spatial cubes are essentially mixed with each other. In such a universe, spatial cubes will have no defined boundaries, and as a result of this, it will be impossible at all to speak of the in terms of past, present and future.

Throughout the book, I refer to the Universe as if the same Existence Algorithm operates in each and every region in space. But this is not unavoidable. The Third Axiom obligates a continuous Existence Algorithm, but it does not require that the exact same algorithm acts on all regions of space. It is possible that—even though it has never been seen in observations—there are regions in our Universe in which a different Existence Algorithm operates. In that region, for example, the Fifth Rule may not hold, or the Seventh Rule operates in an opposite manner. These regions understandably will allow traveling to the physical past.

But, for a moment, let us go with the assumption that the Existence Algorithm in our Universe is not only continuous but also uniform. Is it possible, within the framework of the private Existence Algorithm, to move to the past?

In a certain sense, the answer is yes.

Let us take a pair of twins.

The system of the first twin advances in depth, in the manner of the level of obstruction, which is customary on the surface of the Earth.

The system of the other twin intentionally advances in space with a significant obstruction in depth.

After 100 turns, the second twin leaves the region of the obstruction and returns to the place where he left the first twin.

The second twin will see the Existence from the spatial cube that was his "physical past", 100 turns in depth.

The Existence from his physical past progressed in depth, when he was in an obstruction.

Depending on the initial state, it is possible that the second twin will see not only isolated Existences, but even whole systems that remained as they were in his physical past.

I will give an example to facilitate understanding.

Imagine 100 cars traveling in a column. Each car constitutes the past of the car in front of it. The most distant past of the first car in the column is Car No. 100.

If the first car in the column wants to see its "past", it must step aside and wait for the cars to pass in front of it, until Car No.100 arrives.

Go and learn: when the "Existence" "wants" to see an "Existence" from the previous spatial cubes (its "past"), it must become obstructed and allow the Existence from the previous spatial cubes to advance towards it.

In Classical Physics, when returning back in time and performing a change, we would seemingly encounter one of two outcomes. The first one would be an immediate impact on the present, as illustrated in the movie "Back to the Future". The second: at the moment of the change, the universe should split into two parallel universes. Both cases sound unreasonable.

The explanation of Binary Physics, that the past is a physical plane exactly like the present, and therefore it changes all the time, allows a person to return to the past and do there whatever s/he wants, without any need for difficult tasks, such as splitting the universe or changing entire future.

Nevertheless, Binary Physics requires us to separate from the romantic fantasies of going back in time and meeting our parents… because the past is no less dynamic than the present (or the future), and there is a good chance that—when we go back there—we will not find them at all, or find them with different children…

A Journey to the Future

> *Physical Future: "Physical spatial cubes that are located after the reference cube"*

Motion into the physical future is performed in the framework of Private Binary Physics any time when there is no total obstruction.

The Sixth, Seventh, and Tenth rules require every Existence to eventually move to its physical future.

Exactly like a journey into the past, in General Binary Physics, it is possible to imagine an easy journey into the physical future in a universe with no Fifth Rule (uniformity of the queue). Even though we have no observations to that effect, it is possible that there exist regions in our Universe that do not obey the Fifth Rule, and they will facilitate a journey into the future.

Assuming the Fifth Rule is firm and valid uniformly in our entire Universe, it is still possible to imagine a journey into the future.

As we have seen, the speed of time is different for different systems.

Less obstruction in the system's depth = a system that contains less mass = a more energetic system = the speed of time (rate of change) is faster.

A system which speed of time is faster is an entity that progresses in depth at a faster speed than a system which speed of time is slower. Therefore, as we raise the energy of a system and lower its mass, the faster it will advance in depth and move toward it's physical future at a greater speed, compared to systems with a slower speed of time.

Of course, there is a limitation to the speed of movement in depth (the speed of time), and this limitation is expectedly derived from the cell's response time.

When there is no obstruction in depth, the only limitation on the speed of the Existence's motion in depth is the response time of the cell. Light rays, which are pure energy, proceed in depth at the fastest possible speed, and therefore they perform a faster journey in time in relation to systems that have an abundance of mass, such as human systems.

For example, a human system observes a light ray. A second later, the human system again observes the light ray. The Existence that comprises the human system is (almost) the same Existence that comprised him a second before. But due to the rapid rate of flow of the Existence in the light ray, the Existence that makes up the light ray is the Existence that made up the light ray trillions of spatial cubes earlier... That is, when we take a look around us, we see the same light rays that the dinosaurs see...

In theory, a human system accelerated to the speed of light would fly at the maximal speed of time into the future, zoom past the spatial cubes, and would see thousands of generations of human systems joyful and weeping, living and dying... all this in less time than it takes me to type a single letter of this book... When he would slow down again, and expand and increase his internal obstruction (his mass) back again to the pattern of human systems as is familiar to us, he will find himself far, far away in the physical future...

I will give an example. Picture two rows of people that advance, with the first row advancing quicker than the second. After some time, the first row is obstructed, and then the second row advances faster, and the people in the first row see the people from the second row that they met in their past.

Gliding into the future can be performed through a region with a greater speed of time, such as, for example, near a black hole. Gliding into the past can be performed by being located behind an obstruction.

Restrictions to Time Travel

It is impossible to glide into the future at a faster speed than the speed of the response time of the cell.

You must remember when gliding into the past that the past changes as well.

When gliding into the past, the glider essentially is standing in place (obstructed), and "the past" advances toward him.

When gliding into the future, the glider advances at a faster speed of time than its surroundings, and thus it proceeds in the spatial cubes in relation to the systems with a slower speed of time.

It is possible that the Existence "goes out" from one point of the Big Bang, and converges again at a second point of the Big Bang, at another place in depth and in space, and back again.

It gives an impression that in our Private Binary Physics, in every cycle of the Big Bang and convergence, some of the Existence "disappears" (since a portion of it scatters and will not succeed to return to the process of convergence). Thus, in every such cycle, the concentration of the Existence at the point of the Big Bang will decrease, until the entire Existence will scatter throughout the expanse of the Universe. This is of course only one possibility, and at the moment I do not have an opportunity to determine whether this is the possibility that attains in our Private Binary Physics.

A Solution to the Time Traveler Paradox

In the framework of Private Binary Physics, a system in general, and a human system in particular, cannot move to a previous time cube. This contradicts the Sixth Rule: the symmetry of space-depth & Seventh Rule – Trivial Movement in Depth.

However, just by way of example, let us assume that a human system passed to the previous spatial cube in dimension $d(4)$.

For example, let us assume that in the initial state a human system is located in spatial cube $d(5)10d(4)10$, and it "went back in time" in the next queue to spatial cube $d(5)11d(4)5$. What will be the effect of this move?

It is understood that the move will affect the spatial cube $d(5)11d(4)5$—the spatial cube which the system has reached at this moment. The move is also expected to influence the spatial cube $d(5)11d(4)11$, which is the spatial cube to which the system was supposed to move (it will disappear from it…)

The move will not influence the spatial cube $d(5)11d(4)6$. In the next turn, the Existence of the spatial cube $d(5)11d(4)6$ will pass on to the spatial cube $d(5)12d(4)7$, independent of the values of the cells in the spatial cube $d(5)11d(4)5$.

The Existence of the spatial cube $d(5)11d(4)5$ can influence the spatial cube $d(4)11$ only for 6 turns afterwards—that is, at $d(5)17$.

At this stage, the Existence of d(5)11d(4)11 will already be in the spatial cube d(5)17d(4)17…Even if you kill your father at depth d(5)11d(4)5, it will not influence in any way d(5)11d(4)11, which is what the mind interprets as the "future" that you just left…

In a strange way, your passage to the spatial cube d(5)11d(4)5 will change your past. Since the obstructions that were formed by the Existence in the spatial cube d(5)11d(4)5 influence the spatial cube d(5)11d(4)4….

It should be remembered that the fundamental premise is that the entire sequence of depth that we call past and future are subject to incessant change, just like the present.

If we go back now 10 years in time, not necessarily will we recognize the place where we were 10 years ago, because the "past" is changing all the time.

To facilitate understanding this, let us take, for example, a river. If you take a drop of water from the end of the river, and put it at the beginning of the river, this drop will not affect the entire flow of water in front of it at the moment you have placed it. Assuming the flow of water progresses at a rate of a centimeter per second, and the length of the river is a meter:

(1) The moment that we put the drop in, the flow at the point after the place where we put the drop in the river, will not be influenced at all.
(2) A second later, only one centimeter of the river's length will be affected.
(3) After two seconds, two centimeters of the river's length will be affected.
(4) Only after 100 seconds, the entire river will be affected by the drop.

The flow at the end of the river will not change, except at the point in time in which the drop reaches the end.

One of the reasons why it is difficult to move in time at a speed faster than the average speed of time in your space, is that you are obstructed by the flow of the Existence near your system in both space and depth.

It is just as a drop of water in a river, which wants to move faster than the speed of the flow around it, will encounter resistance. If it would want to move upstream, drops behind it would obstruct it, and if it would want to move downstream, drops in front of it would obstruct it.

Chapter 24
A Brief History
of The Consciousness

Is there a possibility that within our private initial state, and our private Existence Algorithm and after the passage of a sufficient number of turns, that an observer can see an observation of a Consciousness? Yes.

It is important to emphasize—that since we are dealing with physical investigation, the emphasis is on observation of the consciousness, and not on the consciousness itself, however it may be. That is, an observation of the system algorithm of the type that answers to the definition of consciousness. The question of whether the consciousness has a special meaning beyond the very existence of the observation upon it—goes beyond the scope of physics research.

> *Awareness:* "*Algorithm based on the influence of one Existence on other Existences.*"
>
> *Consciousness:* "*A system that produces an output for itself, and responds to that output.*"

Consciousness itself has no value. The system algorithm existed prior to it, and will exist after it. The system algorithm does not require "a consciousness" in order to develop. Furthermore, it is definitely possible to describe system algorithms that are more successful in preserving their conscious-less systems than our system algorithm, which possesses consciousness.

Also, as a physical observation, consciousness is a marginal observation.

At least as far as we know, the Universe has already existed for billions of years, and human consciousness exists, at best, only a few hundred thousand years – a fraction of a percent.

Notwithstanding, consciousness has a very significant part in physics. This is the only known means by which an observer internal to the system can be aware of the existence.

I felt that it would be impossible to write a book describing the physical universe in a general manner, without relating to the instrument which we sense things with.

When considering the consciousness with physical tools, it should not be viewed differently than any other observation.

The consciousness did not suddenly appear *ex nihilo*, and it doesn't have any rules of its own.

The consciousness is a physical observation that results from the same axioms and the same rules from which all other observations are derived. Therefore, it must be explained in this manner.

The consciousness is an algorithmic process at low resolution and stems from the Existence Algorithm.

The beginning of consciousness is awareness.

Awareness is defined by the Existence Algorithm.

Awareness comes into expression already in the First Rule – the

Rule of Autonomy. To determine the value of the cell according to the values of the cells around it, it must be "aware" of the values of those cells.

A simple example of this can be seen in the Third Rule– the Rule of Non-Merging.

If the Existence cannot pass to a neighboring cell that contains another Existence, it must be "aware" of the value of that cell.

Parenthetically, I will note that it is possible in General Binary Physics to describe universes in which there is no awareness. In such universes the First, Third, Eighth, Ninth and Tenth rules cannot be applied. In such universes, Particles will not be aware of the existence of one another. Particles will disappear one into the other, and particles will appear suddenly one out of the other…

The Existence Algorithm contains also the foundation of self-awareness. The particle must "know" that it is the Existence and not Placeholder.

From the awareness of the neighboring cells and awareness of its own value that was determined in the Existence Algorithm, systems at low resolution developed awareness of other systems at low resolution.

I will explain.

On the basic regularity of one Existence that reacts to another Existence, it is possible to derive, at low resolution, the regularity of a system of 4 Existences that react to 4 other Existences.

At low resolution, it is possible to relate to all four Existences as a single system that is aware of the existence of another system that is also composed of 4 Existences.

The more systems develop at lower resolution, the more Existence is used for building these systems, and thus their awareness changed to be more complex, until it becomes what human systems call "consciousness".

I want to say that if we run the Existence Algorithm on a computer for enough turns and set a particular initial state, we can identify—acting as observers external to the system and looking at a sufficiently low resolution—systems that we will define as possessing consciousness.

As we have seen in the chapter on the development of systems, the system algorithm that controls them is an algorithm at low resolution, which succeeds in causing them to survive in the physical reality of scattering of the Existence.

The more the Existence gets scattered in the Universe relative to the Initial State, the more complex and sophisticated system algorithms are required in order to survive.

The evolution of systems algorithms started immediately in the first turn after the Initial State.

At some point, very complex system evolved at very low resolution, that began to respond as a group to the environment.

Here are eleven stages of the development of consciousness that we will see in the evolution of a system algorithm:

1. The awareness of an Existence about another Existence through rules of the private Existence Algorithm.

2. The awareness of a system of the Existence about another system of the Existence—the private Existence Algorithm at low resolution.

3. The development of a "living" system algorithm = "a system algorithm possessing an active mechanism to avoid decay". This active mechanism is an initial reflex. From a physical perspective, a distinguishing feature of a "live" reflex compared to an "inanimate" response is that the main goal of the active reaction is to help the system preserve its existence. When a marble hits another marble, this is a passive response that has no purpose in maintaining the system. When a cat hair bristles, this is a reflex—an active reaction—that its main goal is to help him preserve his system. An algorithm of the type of initial reflex is—an input cell that activates an output cell.

4. Development of a processing and memory layer between the input and the reflex (forming output that is based on processing a number of input sources) — "conditioning reflex". An algorithm of the type of conditional reflex is—an input cell that activates a memory cell that activates an output cell.

5. A more refined processing layer, in which sub system-algorithms like "imagination" (simulator) are developed. "Conditional reflex" develops. Imagination is essentially a part of the processing layer which tries, before acting, to estimate results according to previous experience. An algorithm of imagination is of the type—an input cell that activates a memory cell, that activates another memory cell, that activates an output cell. The innovation here is with the memory cell that activates another memory cell, and not the memory cell that activates directly the output cell.

6. Development of communication between systems—one system is capable of influencing another system by her output. As a result of communication, language is developed.

7. An abstract imagination develops. The difference between

abstract imagination and regular imagination is that regular imagination is based on a memory cells which each one received its data directly from an input cell, and thus it reflects an existing reality. The memory cells of abstract imagination receives its data from other memory cells, and therefore it is able to combine data that does not exist in reality. The algorithm of abstract imagination is of the type: an input cell activates memory cell A. The input cell activates memory cell B. Memory cells A and B activate memory cell C, that now contains data which does not necessarily exist in reality, but rather constitutes a mixture of the information that is in memory cells A and B.

8. The system starts to be influence from its own output. . For example, the system sets off a warning whistle for other systems—it hears it itself and responds to it. The system Coincidentally influence itself by means of its own output. Communication of the system with itself improves the chance of its survival, and thus evolves.

9. The system starts to create dedicated output intended to be received by itself—the system start to talk to itself. A stream of consciousness develops (internal speech). The system develops an initial self-awareness.

10. The system, with its inner speech with itself, identifies its own system as an ultimate goal that must be preserved, and thus sets itself apart from the environment. There is no guiding hand, but rather systems that have randomly formed a system algorithm that identified its own system as an ultimate goal that needs to be preserved, survive more.

11. The system, that already has capabilities of abstract imagination and language, begins a process of conceptualization of its reflexes (fear, hunger, love). A developed self-consciousness is created, based on conceptualizing the

reflexes. The combination of imagination that developed at Stage 5 and the stream of consciousness (internal speech) that developed at Stage 8 enables the development of rational thought—creating internal procedural logic.

Obviously, this is a very simplified explanation, and for every paragraph it is possible to write a separate book. However, my aim here is only to show the feasibility of development a consciousness from an evolutionary process of development of systems that are based on the Existence Algorithm.

If we run the Existence Algorithm over a sufficient number of turns, more and more sophisticated system algorithms will necessarily develop, because only more sophisticated system algorithms will survive in an environment of scattering of the Existence.

Of course, the final level of refining system algorithms depends on two factors, i.e. the definition of a private Existence Algorithm and the initial state.

There will be initial states and Existence Algorithms that will be unconducive to the formation of system algorithms.

On the other hand, there will be initial states and Existence Algorithms that will allow for the formation of sophisticated system algorithms, such as those that we see in the Existence Algorithms of human systems.

It is reasonable to assume that there are initial states and Existence Algorithms in General Binary Physics that allow for the formation of system algorithms, which are many times more sophisticated than those we see in human systems. Without adversely affecting the generality of things and over the passage of enough turns, it is possible that the system algorithms that we see in our Private Universe will become sophisticated beyond recognition, such that the consciousness that is considered today as one of the pinnacles of the achievement of the system algorithm will be seen as an isolated letter…

I want to emphasize the transition from the stage of a system that receives an output of another system (Stage 7), to the stage of a system that begins to receive its own accidental output (Stage 8), up to the stage where the system creates a dedicated output for itself (Stage 9) – the stage of consciousness.

These transitions are the critical stages in the development of consciousness. Once the system begins to receive its own accidental output at Stage 8, it is forced to identify itself as a source of an output (as it learnt at Stage 7 to distinguish the source of the output of other systems). If you want to put your finger on the exact moment when consciousness was born, you need to point to the moment when the system was forced to identify itself as the source of its own output.

From this point, it is a short distance to Stage 9, when the system begins to produce a dedicated output for itself. This is the beginning of the stream of consciousness – internal speech. When a system produces a warning signal upon receiving a particular input, and reacts by itself to this output, it essentially talks to itself. Of course, it is a far cry from here to the inner speech that is familiar to us human systems nowadays. But the evolutionary processes that systems undergo definitely prove that from the moment when a particular algorithm that assists in the survival of the system has been formed it will be required to be refined to the point of being unrecognizable (like from a cell that is sensitive to the human eye).

At Stage 9, the stream of consciousness starts to develop at an increasingly rapid rate. The system algorithm starts to create more and more self outputs, which assist in preserving the system that contains it. Within the system, "a small news station" starts to operate, that creates relevant self output for survival. Suddenly, a structural sub system algorithm is formed, that constantly creates outputs for the system: "Warning—a lion", "a beautiful girl", "there is food over there" *etc.*

As language gradually becomes more sophisticated, so does the internal speech. A process of conceptualizing reflexes into more abstract elements—"I am afraid", "I love", "I'm having fun"—is being formed.

The ultimate refinement of the inner speech is done by a sort of input-output "ping-pong" conversation of the system with itself:
Input / Output – "I am hungry"
Input / Output – "I will hunt"
Input / Output – "It's dangerous"
Input / Output – "Maybe I'll go fishing"

In fact, the consciousness nowadays is composed by constantly mixing input from external systems and internal input that the system produces for its self. When the central processing system focuses on the external input, the internal input is "weakened", and *vice versa*. When the stream of consciousness feels that there is a decrease in the external input, the internal input is amplified, and the "news station" will report to us unceasingly: "She loves me / She does not love me…"

Consciousness and Will

Consciousness (and awareness) is not related to Will. I can be without any control over my desires and/or my actions, but still possess a consciousness. Human systems confuse the two concepts – Consciousness and Will, and especially Consciousness and *Free* Will. They wrap the concepts together as if it was a single concept.

There is no contradiction between a conscious process and a deterministic process. A process can be both conscious and deterministic. You can understand what you are going through, even if you are not in control of what you are going through.

When I come up with an idea that I want to go to a shopping mall, I am aware of this thought, whether we assume that the desire is authentically mine or implanted into me by an alien, and I have no control over it.

Nor is there any difference between awareness of a process about another process and its awareness of its own process.

The way in which I am aware that Obama went to the mall is exactly the same as my awareness that I went to the mall. It is absolutely clear that my awareness that Obama went to the mall is unrelated in any way to my desire.

Another example: When I fly on a plane, my control over my body's movement is limited. My body is largely controlled by the movement of the plane. Yet my consciousness still exists and can produce an output of the story about my flying in an airplane for me…

I can be in total lack of control and yet still possess consciousness. Since every statement I make (which I call "my decision") is derived from the physical system of my brain, which is derived from the deterministic Existence Algorithm – I have no free will. But I certainly have consciousness.

Is There Life on Other Planets?

My grandfather used to say that in order to find something, first it is worth knowing what you are looking for. I admit that I have not managed to think about it yet (it is the next thing I intend to think about, after I am done thinking about how many changes can a system undergo until it ceases to be the same system…)…

In any event, it seems to me that before we start searching extraterrestrial life, it is worthwhile for us to understand exactly what is the "life" we are searching for…

Binary Physics defines "life" as follows:

> *Life: "A sub system algorithm that actively responds to input, which harms the system, by means of output that prevents its decomposition"*

A system algorithm at low resolution is infinitely more complex than the Existence Algorithm. Therefore, you can consider it to be built of sub system algorithms, each having a more specific role in the overall regularity of the system.

For example, when looking at the system algorithm of human systems, it is possible to say that it is composed of many sub system algorithms – the sub system algorithm of emotions, of breathing, of the heart, and so on...

The more we increase in resolution, the more each system algorithm is broken down to sub system algorithms, until we reach the Existence Algorithm.

If the system algorithm of a particular system contains a sub system algorithm that is capable of actively reacting to a change in the environment, and thus to prevent the decomposition of the system, this is a "living" system. Easy and simple.

I emphasize the word "actively" in the definition of life. The atom also has a system algorithm. But its system algorithm does not contain an active element of its preservation. As long as the input around it allows for it, it continues to maintain its system, in accordance with the Existence Algorithm. Once the input around it has changed, the atom will decay, and we will not see any active response on its part to prevent the decay. To the atom's credit, it can be said that it is a system at a relatively high resolution, and therefore the system algorithm that operates it is relatively simple, and thus it does not possess sufficient complexity to contain a sub system algorithm to actively prevent its decay.

A good definition from the field of biology for a sub system algorithm that contains an active response against decomposition is the reflex.

Of course, we are not speaking of a reflex of any sort of response to some action on the system. To distinguish between a reflex of a "living" and a "dead" system, its response must help the system survive, even if it is very simple. I call this a "**survival reflex**".

The basis of life is an isolated survival reflex. Every system that has a sub system algorithm that contains even a single survival reflex—is alive.

Of course, just as there are systems at high resolution that are constructed from a (relatively) limited number of the Existence particles, such as the electron, and systems at low resolution, such as human systems, that are built from a (relatively) very high quantity of the Existence, so is there life based on one survival reflex, and life that constitutes a superposition of a huge number of survival reflexes, such as human systems.

A human system is a good example of a system that actively responds to the environment and by doing so, avoids its decomposition. This is a system that contains a tremendous amount of survival reflexes. When the consciousness considers the combined action of the various reflexes at low resolution, it finds it difficult to identify each and every reflex. The consciousness sees the human survival sub system algorithm system as a one very complex regularity.. There are many examples of survival reflexes that constitute a part of the human survival sub system algorithm: the sleep reflex, the hunger reflex, the breathing reflex, the metabolic reflex, the pleasure reflex, the love reflex… Of course, every reflex that I have just mentioned is composed of many sub reflexes (sub system algorithm), and if we increase in resolution, we will again arrive at the Existence Algorithm…

Another key example of a survival reflex in human systems to prevent decomposition of its system is the sub system algorithm of reproduction. This sub system algorithm recreates the systems (in light of the genetic reproductive system), and thus enables older systems to be destroyed without damaging the very existence of the systems (since new systems arise in their place).

As a side note, I should mention that the sub system algorithm of preservation through reproduction is a nice, but inefficient and primitive system algorithm for preserving systems. It is possible to imagine methods that are many times more efficient. Why ruin the old system every time to engender a new and vulnerable one, instead of constantly improving the existing system? To rebuild the system for at least 20 years, of which 16 years are invested in teaching it knowledge that already existed in the systems that created it… A system algorithm, which simply knows how to improve itself without a reproductive procedure, is a system algorithm that is many times more successful, in terms of its ability to survive.

I believe that soon we shall see for example robot systems, which do not contain a sub algorithm of reproduction, and yet manage quite successfully to preserve their system and even improve it.

What is certain is that as soon as human systems will have the possibility of forming people without birth, they will do so. That would be the twilight of the emotions of love, intimacy, and love for one's children.

And another remark: if we are already speaking of birth… then a word about old age…

Aging is not a necessity of the system. It is possible to see many stable systems that will never change, unless there is some external intervention. It is also possible to describe durable systems that even in a state of attempting to intervene, no change will occur. Alongside them, it is possible to observe systems that suddenly decay, without any gradual process. As a result, the system algorithm, that decay the system gradually in a process called "aging", constitutes only one out of many possibilities.

A system sub-algorithm of Decay in a process of "aging" has developed over the course of evolution in certain systems as a method for increasing the chances of survival. A system sub-algorithm for aging was formed in order to improve the chances of survival of new (and improved) systems by means of a process of self-destruction of older systems. Thus resources are freed up for the benefit of new systems. The system sub-algorithm of aging is a derivative of the system algorithm of reproduction. If there is a system algorithm (that is more successful) for improving an existing system—there is no need for a mechanism to destroy it. Therefore, from the moment that human systems will succeed in improving their system without reproduction, the need for a sub-algorithm of aging will also become superfluous. To be victorious over aging does not mean to invent a new reality, but rather only to neutralize an obsolete system sub-algorithm.

Of course, life is a mental concept, and not a physical concept. There really is no life, just like there really is no system and there really is no system Algorithm. There is the Existence and the Existence Algorithm, and at low resolution, our consciousness imagines that certain systems have a regularity of their own—and a regularity of a certain type that knows how to actively respond to a change in the environment to prevent decomposition is called—life.

There is a familiar phrase "fighting for his life". This phrase is attributed to a human system that is on the verge between being a system that the consciousness interprets as alive and a system interpreted as dead.

I think that, along this boundary between life and death, it is worth looking for what our consciousness pays attention - The struggle for life. The struggle against the decomposition of the system.

What distinguishes a "living" human system from a "dead" one is the sub algorithm that allows it to actively maintain its system. Once this sub algorithm stops operating, the human system is dead.

Since the system algorithm of the human system is complex, so is its sub algorithm that is called "life", which is composed of many sub algorithms—the central brain, the nervous system, and local reflexes…

If the central brain is dead, it is customary to refer to the entire human system as dead. Actually, in such a situation, a human system cannot actively respond to the environment to prevent its decomposition. But, if we increase in resolution, it is possible to see also within the dead human system sub system that are still alive, such as cells systems that still actively respond to the environment to prevent their decomposition.

If a human system is in a vegetative state, but part of its central brain is still functioning—that is, it engages in dynamic activities to preserve its system—it is possible to say that that part of its brain is still alive, even if the activities do not go into effect (to the muscles, for instance). When there is still an active system algorithm in the brain to preserve the system, but there is a disconnection in transferring the output to other sections of the body, is it possible to define the human system in general as alive? This is a question of a mental definition, not a physical question. From a philosophical standpoint, it is possible to argue that since the brain belongs to a particular body of a human system, then if the brain is "alive", we should treat the entire body as "alive". From a physical standpoint, the living section is only the part where the active sub system algorithm is fulfilled and influences it to preserve its system.

From a physical standpoint, it is possible to relate, for example, to the country system as a living system. The system of a country is indeed built at a resolution lower than a human system, but the system algorithm of a country contains many sub system algorithms that actively assist in preserving its system (firefighters, army, police, courts of rule, *etc.*). From the standpoint of physics, there is no difference between the vitality of a human system to the vitality of a country or company.

It is also possible to explore the topic of defining life from the opposite direction as well: a living thing is something that can be killed.

Who can be killed? It is possible to kill only someone who wants to live. Wanting to live = it has an active sub system algorithm to preserve its system.

Therefore, if I will now program a creature on the computer, that is composed of 3 pixels, and, as a part of its system algorithm I will program an active sub system algorithm that will help it preserve its system from being erased by the mouse—it will be alive. For example, I will program it so that every time that the mouse cursor approaches it, it will run away from the cursor.

I know that it is difficult to accept this at first, because I gave an example of a very simple sub system algorithm for preserving its system.

But imagine a very complex program with artificial intelligence that includes a very elaborate sub system algorithm for preserving its system, which—when you try to turn it off—starts fighting you back in very creative and smart ways… I think that then you would agree with me that you are dealing with something quite alive… something that very much does not want you to kill it!

My three-pixel little creature, simply possesses a very primitive system algorithm, that only knows how to move it to the right and left when you, out your great cruelty, try to delete it with the mouse… then you show contempt for its life… but if I would refine the sub system algorithm of my cute little creation in such a way that every time that you would try to kill it, it would be able to organize and operate an arsenal of nuclear weapons in order to save itself… you would try to preserve its life much more seriously.

And one final example: If we will find a rock on Mars that has a system algorithm that determined that when we try to hit it with a hammer, it moves to the side to avoid getting hit, this rock would be—from a physical standpoint—alive.

When a common human system thinks about life on other planets, it is not thinking about "life" in its physical definition, but it is influenced by the "similar-to-me" paradigm, and searches for a system possessing sub system algorithms similar to its own. A central sub system algorithm of human systems, which actively helps them survive and thus constitutes a part of the sub algorithm of life, is the sub algorithm of emotions. Therefore, human systems search the Universe for other systems possessing a "romantic" sub system algorithm… like those that possess emotions… aliens that want to rule over us are indeed evil, but very similar to us… they possess a romantic behavior… they have emotions… they want to rule over us…

But living things with a romantic sub system algorithm are a very specific case in the group of "living things".

We need to remember that the physical term of human systems is very short, and therefore chances are great that we will see a planet before or after the appearance of human-like systems. If the aim is to find life, we have to scan the Universe and search for life in general, and not a particular instance of life, which is our particular case, as human systems.

We must search for a rock or a system with a reflex that allows it to preserve itself by actively responding to the environment in order to avoid its deterioration…

In short, someone who searches for life must look for a moving rock, and not a green alien…

I believe that if we will search well, we will find quite a few such as those…

Life and Resolution

It is possible that some of the reasons that we do not see "life" on other planets is that this life exists at a resolution different from ours.

A conscious, "living" physical system, which exists at a resolution lower than ours, will compress more cells into each of its "seconds" (each of its seconds will be longer), and therefore the way in which it will grasp physical reality will be different than ours. If the resolution of that physical system will be lower or higher than ours, the human consciousness will not be able to grasp it.

In fact, it is possible that even on the Earth there exist living physical systems, but we are not aware of them, because they are at a different resolution.

Continuation of the Development of Consciousness

Over the last hundreds of thousands of years, the system algorithm of human systems has developed and very advanced reflexes have formed in it that are called "emotions". The emotions are a part of the group of sub system algorithms of "living" systems.

What drives the development of our system is only its survival in the Universe in which the Existence dissipates.

Therefore, we will see that the system algorithm improves its methods of survival more and more.

For example, we will invent new medicines and succeed in preventing our systems from deteriorating due to illness and aging.

At the same time, competing system algorithms will continue to develop – like system algorithms of artificial intelligence that will live in machines.

Happiness

> *Parasitic sub system algorithm: "A mental concept that indicates a sub system algorithm that does not assist and/or damage the preservation of the system"*

The system algorithm develops from the Initial State in an evolutionary manner.

It is possible to see successful system algorithms that preserve their system well, for example, the system algorithm of the atom and the system algorithm of the human system.

Of course, alongside them developed system algorithms that did not succeed in preserving their system, animals that became extinct or elements that disappeared.

Since the system algorithm develops and changes, "successful" and "less successful" system algorithms are being formed all the time.

Sometimes a sub system algorithm develops within the system algorithm, which creates regularity in the system that does not assist its survival, or worse, brings about its extinction.

Such a sub system algorithm is called a parasitic sub system algorithm.

A parasitic sub system algorithm can be more or less "violent". The less "violent" one has lesser influence on the system and thus a chance that it will bring about the extinction of its system is less.

A parasitic sub system algorithm that survives must be violent to such a level that it does not lead to the extinction of its system, and thus its own extinction.

It is possible that at some point a parasitic algorithm takes control of the main system algorithm, just as a parasite takes control over its host in the Animal Kingdom. One of the likely consequences of such a takeover is the non-survival (destruction) of the carrier system of the parasitic system algorithm that rules over it.

As long as the parasitic sub algorithm does not cause the extinction of its system, it will continue to develop evolutionarily in parallel to the evolutionary development of the main system algorithm.

Because an over-dominant parasitic system algorithm may destroy its system, and thus destroy itself, it is reasonable to assume that we cannot see a high frequency of super-dominant parasitic system algorithms in observations.

We will find more examples of non-aggressive parasitic system algorithms that influence, but still live in peace with, the main system algorithm of the system in which they dwell. The reason for that is simple. It is possible to find algorithms that survive, whereas algorithms that do not survive are impossible to find.

> *Happiness is an example of a parasitic sub system algorithm.*

Pleasure, as opposed to happiness, is a normal sub system algorithm. It directs the system to perform activities that are necessary for its survival.

The parasitic sub system algorithm of happiness directs the system to produce pleasure for pleasure's sake, in a way that is unrelated to its survival.

For example, the parasitic algorithm of happiness causes a human system to go on "vacation".

By comparison, the concept of "vacation" does not exist in the animal world. I do not recall any observation of my two dogs going on vacation…

Human systems, waste enormous energies, which are not utilized for survival of the system, due to their desire to enjoy a vacation.

A distinction should be made, of course, between the concept of vacation and the concept of rest, which certainly constitutes a part of the system algorithm.

There is no survival benefit in vacationing, apart from "satisfaction" of the parasitic sub system algorithm of happiness.

There are even human systems whose parasitic sub algorithm of happiness makes them take "extreme vacations" such as skiing or parachuting… things which certainly do not assist their system algorithm maintain their system…

A human system whose parasitic sub algorithm of happiness will be "violent" to such an extent that it will be the leading tone in its system algorithm, will become extinct.

In this context, I recall an observation in an experiment with rats, when they stimulated their pleasure center every time they pressed on a pedal, and from that moment onward they stopped eating or performing any other activity, and they pressed on the pedal until they died…

Of course, as a human system that contains the parasitic sub system algorithm of happiness I am an avid supporter of maximizing the general happiness, and even try tirelessly to maximize my own personal happiness…

But as a physicist, I am not sure that the strategy of maximizing happiness is the best advice that I would give to my fellow human systems if they want to survive...

Are we likely to be happier in the future?

Imagine a situation, which is very likely by the way, developing new pill that neutralizes the parasitic sub system algorithm of happiness and the person who takes it does not feel the need to be happy.

I can imagine that a human system that will be treated with this pill will be a good worker. He will not be depressed because his girlfriend left him, and he will not want to leave his job and seek happiness in an ashram. He will not take a week of vacation, and he will order a basic lunch to his office in order to satisfy his energy needs, and will not go for a stroll for an hour, at the expense of working hours, to a new restaurant that just opened...

At first, this pill will be a curiosity. Only crazy workaholics would use it.

Gradually, as its use would increase, an exponential process will begin.

When most of the employees in Company A will start taking the pill and their work productivity will increase significantly, tremendous pressure will begin in the Board of Directors of Company B for their employees to take the medicine as well as a condition to continue working there...

At the same time, system algorithms of artificial intelligences are expected to develop with no built-in parasitic sub system algorithm of happiness, because it is clear that adding it would only harm their functioning.

Also, the artificial intelligences systems that are indifferent to happiness will compete with human systems, and will compel them to make every effort to get rid of the parasitic sub system algorithm that disturbs them from focusing on their survival...

In the human systems' mythology, they are so central in their own eyes that in any future scenarios that they envision, they tell themselves the story that their system algorithm tells them, rather than the physical story that comes from the observations. In their system algorithm story, which is concerned that their system will survive, they will become immortal. And their parasitic sub system algorithm will bother to add – happy immortals...

However, when looking at the envisioned future of the human systems from a physical aspect, while discarding the effect of the parasitic sub system algorithm of happiness, it is possible to say that human systems are likely to be much more stable and durable and possibly even immune to disease and death, but it is not really certain that they will be happier.

There is only one clear line of physical history - the story of the survival of the systems. Ie, turning more stability and more durable. It seems that over time there will be the moment that the parasitic algorithm of happiness would become a disadvantage for survival, and disappear. This assumption fits very well with the observation that we do not see in the universe "happy systems", except those known to us from Earth. It may be that The parasitic algorithm of happiness appears briefly in certain circumstances and for evolutionary reasons is gone. Therefore, it may be that the planets we are observed are before or after the very short time, in physical Concepts, which in the parasitic algorithm of happiness develops.

Since we stand at the end of a revolution, in light of deciphering the genetic code and development of artificial intelligence, I would not be surprised if it will be possible to say about our generation that it is the last happy generation... you are allowed to smile... We Win.

To end this discussion on an optimistic note, I will say that there is no such obligation to be derived from the General or Private Binary Physics, and the parasitic sub system algorithm of happiness can continue to thrive.

It is almost impossible to foresee a direction in which the system algorithm will develop (and as a result, the way in which its sub system algorithm will develop…). For a complete view of the way of its development, it is necessary to know the location of each and every Existence in our queue, and in all the subsequent queues… this is something that is beyond our capabilities…

Chapter 25
The Will

"Will" is derived from the Existence Algorithm, i.e., from the rules of Private Binary Physics.

Without the Existence Algorithm, there is no will.

At the highest—fundamental—resolution, will is derived directly from the Existence Algorithm, i.e., the rules of Private Binary Physics.

At low resolution in general, and in particular at the resolution of human systems, Will appears to us as a very complex phenomenon. This is because the perspective at low resolution is a superposition of thousands of small "wills" of the Existence.

Will is derived from the following four rules of motion of Private Binary Physics:

The Seventh Rule – Trivial motion in depth
The Eighth Rule – Inertia
The Ninth Rule – Collision
The Tenth Rule – Gravity – Bypassing in depth

These Rules can be viewed from two perspectives.

The first is that of an observer external to the system who says:

"There is a rule that states that the Existence must move to another cell."

The second is that of an internal observer within the system who is composed of the same Existence and says: "I want to move to another cell".

Of course, the second point of view requires, as a precondition, a system that is capable of expressing that will (that is, a conscious system). But—and this is an important 'but'—the fact that the will exists or not, does not depend on whether it is expressed. Just like a tree that falls in the forest makes a noise even if no human system hears it.

As soon as we state a rule that requires a particular action – will is born.

Every Existence in the Universe possesses a will.

Only a complex system of the Existence, at low resolution and possessing a consciousness, is capable of feeling a will, understanding that there is a will, and expressing that will.

But will, by itself, is inherent in any Existence in the system.

When, by the Seventh Rule, over the course of a turn, an Existence must advance to a cell in depth, it is possible to describe this action in the language of "will". The Existence wants to advance in depth. The Existence wants to pass from one cell to another cell.

In the absence of rules that determine that, over the course of a turn, the Existence must perform some action, it is not possible to speak of will. An Existence upon which no Existence Algorithm acts, and which remains static as the queue transpires, is an Existence lacking any "will". In General Binary Physics, it is certainly possible to describe universes lacking any "will".

We are accustomed to use the word "will" in the context of complex systems, first and foremost human systems. When a human system moves to the right, we tend to describe this physical operation by saying, "Joseph wanted to move to the right". But Joseph wanted to move to the right only because the system Algorithm that operates on him caused him to move to the right.

From a physical standpoint, water wants to flow downwards, just like Joseph wants to move to the right. There is no physical difference between the two desires.

Nevertheless, we human systems feel that we desire, whereas water does not feel this desire.

Thus, the very feeling, or lack thereof, about a certain thing, does not affect its existence. The fact that water does not feel a desire does not mean that its desire does not exist.

Unequivocally, water has a desire to flow downward, and this will is a physical will, completely identical to my desire to move to the right.

To summarize, the will is not related to awareness, and, as a derivative of this, to consciousness. There can be a will without awareness or consciousness.

Human systems developed consciousness, and as a derivative of it, they are capable of describing the process that transpires to them. To the physical process that causes me to move to the right, I call "will". When my awareness contemplates water, it calls this physical process by its more correct name, "rule".

The reason that our consciousness describes the rule that operates on the system in which it exists as "will", whereas the rule that operates directly on another non-consciousness system by the name "rule", comes from the additional sensation that the consciousness has regarding its system—the sensation of control.

Free Will

"Man is born free, and everywhere he is in chains" – Jean-Jacques Rousseau.

Human systems sense free will.

Binary Physics defines free will as following:

> *Free Will: "The mental concept that denotes an autonomic ability of a system to influence her output"*

Before we expand on the analysis of free will, we will define two more terms.

> *System Input: "A mental concept denoting the impact of the environment on the system"*

> *System Output: "A mental concept denoting the impact of the system on the environment"*

The First Rule—the Rule of Autonomy—states that change in the position of the Existence is determined autonomously according to the Existence around it. Therefore, you can refer to the Existence around a particular Existence as input, and refer to the change in that particular Existence as output. At lower resolutions, the same exact principle operates, just on the level of systems.

The essence of free will is the ability of a system to determine its output autonomously.

For example, a human system that can determine what will be his output—to kill or not to kill—possesses free will.

The output of the system is affected by two factors: input and the system algorithm.

That is, in order to have free will, the system must be capable of controlling its input, and/or its system algorithm.

Control of the system algorithm: There is no doubt that a system that control of its system algorithm is the embodiment of free will. The problem is that such control contradicts the Third Axiom, which declares the existence of a continuous Existence Algorithm. A system internal to the system is unable to change the system algorithm that operates it, because the system algorithm derived from the Existence Algorithm that can't be changed. Therefore, only a system external to the universe system can change the Existence Algorithm that operates it. If "God" would have taken the form of a human system and entered the universe system—He is not subject at all to the Existence Algorithm, or if He would be able to change the Existence Algorithm that operates Him by the power of His thought—despite the fact that human systems could see His system within the universe system, He would not constitute a part of the universe system. Physics, as a science that studies regularity, is incapable of studying regularity that changes into irregularly. Once we have determined that a system can change its system algorithm based on something that is not the Existence Algorithm, we have negated the reality of the Existence Algorithm, and thus the regularity of the entire universe system.

Controlling input: When a system is capable of controlling the input, it is capable of controlling its output, even in a situation of a continuous and unchanging Existence Algorithm. That is, if a human system is capable of taking care that input of another human system who is cursing him should not be heard by him, it is reasonable to assume that he will succeed in preventing his system from producing output in the form of murdering the human system doing the cursing.

Even the control of input—unfortunately, as someone who supports and desires free will—is contrary to the axioms.

Already at the Initial State, the location of each and every Existence—and as a derivative of this, of each and every system—in the universe system was determined. From there onwards, the change in location of systems has been developing according to the Existence Algorithm. In other words, there is no system with autonomous ability to determine what will be its input. A system that is capable of controlling its input essentially eliminates the regularity of the entire universe system, and therefore cannot be an internal system within the universe system.

In Private Binary Physics there is a more significant reason for the lack of "free will", as human systems define it. The reason for this is that the future influences human systems no less than their past. The obstruction in the next depth overwhelmingly affects the change of the system in the present depth. How is it possible to say about a human system that he has moved his hand to the right of his own free will, when he has done so only because the place is obstructed on his left side in depth in his future?

Thus, in the absence of free will, why do not we cancel the rule "Thou shalt not kill"? What is the point of sentencing a man for murder, when it is absolutely clear that from a physical standpoint the act was not up to him?

The answer is concealed in the question. Because there is no free will. Despite the illusion that we feel that we are capable of choosing to annul the rule "Thou shalt not murder", we really are not. Just as we have free will to choose to commit suicide and we cannot implement it (because the system algorithm of the absolute majority of us causes us to fear it, and tells us that this is not a good idea…), or as we have free will to leave the partner that we love (but the system algorithm does not allow us, because we love that person…).

If the line of development of our system algorithm will allow us eventually to annul the rule, not to murder, and we will drop atomic bombs on each other—okay. Then that will be what has supposed to happen, and there would be no other possibility. By the way, since General Binary Physics, in the Seventh and Eighth dimensions contains all the possible initial states and Existence Algorithms, this is one of the scenarios of General Binary Physics. Thus, from the aspect of pure physics, this presents no problem.

In any event, my private system algorithm, recommends all of us not to annul any rule in general, and the rule "Thou shalt not kill" in particular… I prefer my own Private Binary Physics with my system intact within it… thank you.

In order to calm us, there are chances that knowledge of the lack of free will shall not cause us to annul the rule not to commit murder. Our system algorithm has evolved in such a way that it will not allow us. Just as the parasitic sub system algorithm of happiness will not allow us to give it up so easily...

Whence does the sensation of free will come?

From a physical standpoint, human systems are independent systems. This rule stems from the First Rule – The Rule of Autonomy. Our system gets "its decisions" independently, on the basis of the input that surrounds it. It does not make the "decision" as to its mode of operation from another source, or from some "central consciousness". Therefore, the feeling of our system that it "wants" is certainly understandable. If our consciousness analyzes the observations, it will reach the conclusion that its system has made the decision to move to the right, only because there is no other entity that has made that decision. Our consciousness is quite correct. Its own system has made that decision independently. Therefore, the physical action that has been performed certainly reflects the individual will of its system. But our consciousness makes a slight mistake in interpretation. From the fact that its system possesses an independent "will", it is impossible to conclude that there is also free will.

This interpretive error is due to our consciousness's limited observational ability of its input, and an erroneous conclusion—at low resolution—that it has the potential to control the input, or the ability to control the system algorithm that runs it.

Since the Fourth Axiom states that the Initial State of our system is determined for all cells without exception, then the control of input does not exist in our system. All input is due to the input of the previous turn in the queue, and so on until the first turn and the Initial State.

But our consciousness is unable to understand or observe the entire Universe through all the turns simultaneously, and therefore it interprets the local input that reaches it as the input that it has chosen.

And thus, because our consciousness feels (and it is correct) that an internal will is inherent in its system, and in addition it feels (and here it is mistaken) that it has control over its input and/or on its system algorithm, it arrives at erroneous interpretation that it has free will.

Let us summarize:

> *At the fundamental resolution, will is the Existence Algorithm that requires the Existence to move to a specific cell.*
>
> *At low resolution, the consciousness senses free will, due to the false sense of control over its input and its system algorithm.*
>
> *Awareness does not depend on will, and will does not*

depend on awareness.

To conclude the discussion of free will on an optimistic note: We have seen that there is consciousness and there is will. All right, there is no free will. There is no reason to be upset over it. This knowledge by itself does not impinge on the experience. Just as the knowledge that the flower that we see is only a system of Existence, does not impinge on my experience of the flower.

Physical Will Versus Conscious Will

Physical will is the will that exists in every Existence at the fundamental resolution, to fulfill the rules of the Existence Algorithm.

Just as we decrease in resolution, it is possible to imagine systems that are more and more complex and system algorithms that are more and more complex, so it is also possible to imagine more and more complex wills.

For example, a static Existence wants to fulfill the Seventh Rule and advance to its next immediate level of depth. When we look at low resolution at a very large quantity of the Existence, which we imagine as a system, we will see the entire system as "wanting" to advance to its next immediate level of depth, and, due to this and being obstructed, it will respond by movement in space.

As systems evolved, and it became possible to look at them at lower resolution, so it became possible to look at their desires at a lower resolution. That is, to compress many small desires to an illusion of a greater and more complex desire.

I will explain by way of example.

Axioms -

Assume a two-dimensional universe with live creatures called "the Flat Ones".

A Flat One is a linear, one-dimensional creature composed of the Existence (in the form of a snake).

The Existence Algorithm of the universe:

The Flat Ones move in the universe, one cell in each turn, in the direction in which they moved in the previous turn.

The first direction of motion of the Flat Ones is determined in the initial state.

The Flat Ones change their direction to the right either when they encounter another Flat One or the boundaries of the universe.

When a Flat One encounters the end of another Flat one, a battle ensues, in which one can take the Existence from him or give it to him. The chance each time to take or give is according to the relative size of the Flat One. For example, in an encounter between a Flat One of 10 and a Flat One of 5, the first one has double the chance of taking the Existence than the second one.

Flat Ones that are born long in the initial state are Flat Ones that are more suited to survive in the environment of the Existence Algorithm that was determined above.

The expected end of the process is the creation of a single Flat One, who will swallow all the other Flat Ones (or a number of Flat Ones, if they will get stuck without any ability to move…).

An observer, who sees the Flat Ones at the fundamental resolution, will say that the will of the Flat Ones is to fulfill the rules of the

Existence Algorithm.

An observer at low resolution will see a group of violent Flat Ones who have an obsession to become as long as possible. This observer will say that the will of the Flat Ones is to be as long as possible.

A Flat One itself has no desire to be long or short. The illusion of the desire to become long is an illusion that is created as a result of observing the universe at low resolution. The will to be "long" is a second-order will, that is derived from the specific character of the Existence Algorithm.

The Existence Algorithm is a first-order will.

The complex desires that we experience as human systems stem from the low resolution, at which our consciousness views the Universe.

If we develop an external algorithm that will observe the system and try to describe what the Flat Ones are doing, it is very reasonable that its output will be: 'They desire to be as long as possible.'

This answer will be correct.

But the desire of the Flat Ones to be as long as possible is a second-order desire.

A Flat One for whom the system Algorithm has adopted a strategy that says "Do not steal", will not survive.

It is reasonable to assume that a Flat One whom the system Algorithm has adopted a strategy that says "Always steal", will survive longer.

A Flat One that the system Algorithm has adopted a strategy that says "Always steal from Flat Ones that are smaller than you" will survive even longer.

If there is a certain percentage of Flat Ones whose strategy allows cooperation (for example, in a rule such as: "If there is a Flat One next to you, identical to you in size, join up with him"), their chances for survival increase dramatically as well.

History is written by the victors.

Therefore, if the Flat Ones that survived were given the possibility, at an advanced stage of the game, of telling us what they want—they would tell us that they want to steal from the Flat Ones that are smaller than they are, or to cooperate with the Flat Ones that are the same size as them. It is reasonable to assume that, at an advanced stage, we will not find Flat Ones that will tell us that they don't want to steal… for the simple reason that the Flat Ones that "thought" so would not survive…

In conclusion: In our Private Binary Physics, nothing stands in the way of the will, except for an obstruction.

Chapter 26
Being, Feeling and Experience

Physics studies the regularities of existence.

The only question that physics can answer is: In a particular initial state, and, given a particular Existence Algorithm, after a certain number of turns, can an observer, external or internal to the system, see a particular pattern of Existence within the system?

This is a question of being—does the particular pattern of Existence exist or not?

Our consciousness interprets the Universe in three ways:

1. Being (existence).

2. Awareness (feeling).

3. Experience.

Physics deals with being and also with awareness. Both are physical entities. This is distinct from Experience, which is a non-physical entity.

Being is an axiomatic essence. It is already derived from the First Axiom. Being exists even if there is no one aware of it or experiencing it. It is not dependent on anything. Since being is directly derived from the axioms, it is an essence that belongs to General Binary Physics.

The tree exists.

Awareness (Feeling) is derived from the First, Third, Eighth, Ninth and Tenth rules of private Binary Physics. Therefore, awareness does not necessarily exist in every universe. There can be universes that are devoid of awareness and feeling. For example, if the Existence can enter a cell that already contains the Existence, certainly there is no need that it must be aware of it or feel it. On the other hand, if the Third Rule is fulfilled—the Existence by definition is aware and feels the Existence that is next to it, and is prevented from running it over...

This is awareness and feeling at the fundamental resolution. At lower resolutions, that are made up of a huge superposition of interactions of the fundamental rules, there are feelings as those, which we are familiar with.

I feel a tree. See a tree. Taste a tree. Smell a tree. Hear a tree. I am obstructed by a tree.

Experience is not a physical entity. I do not need to experience a tree in order to be aware of it or to feel it, and certainly I do not need to experience a tree in order that it will exist. And still, observations show that I experience a tree. Why is it necessary to have the experience of the tree itself? In order that there will be a universe, with all that is in it, being is sufficient. In order to create a bridge between the consciousness and being, awareness is sufficient. What is the essence of Experience? There is no physical explanation.

Of course, I am speaking of experience itself, and distinguish it not only from the senses, but also from emotions such as love or pain, that serve as reflexes at low resolution that provide motivation for action. The emotions definitely can be explained from a physical standpoint.

Experience is also a unique physical observation. I can feel it only within my internal system. There is no possibility of seeing it come into expression in any other observation. When I observe another system, the observation is of being, of being aware of it, and of feeling. There is no observation, according to which I will be able to determine that the other system is undergoing an experience. If the other system will tell me that it is undergoing an experience, the observation that I will have is an observation of feeling: I will hear the story according to which that system has an experience. But I will not have any direct observation of the experience.

In terms of experience, although there is no free will and existence is deterministic, certainly there may be different experiences.

For example, two human systems in a water park enter a machine that paralyzes their bodies and projects a virtual reality movie to their brains, such that the human systems feel themselves an integral part of the movie. Throughout the movie, they have no influence on it, and they are not able to stop it or change it. From the moment that they enter the machine, they lose (by abstraction) their free will, and they are subject to deterministic rules. Nonetheless, they are still expected to enjoy the movie very much (or suffer through it…). We see here that experience, like awareness and consciousness, is not affected by the fact its being lacks free will and its deterministic.

A piece of advice: Because the experience is the only thing that is left for us human systems in our deterministic Universe, therefore the only advice I can give you is that you should experience and enjoy, as much as possible, the journey in the Universe that has been organized for you and is not dependent upon you in any way—whatever the experiences may be.

Meaning

Experience – " Meaningful awareness (feeling)"

A qualification: I use the word "experience" and speak on behalf of all of humanity, even though from a physical standpoint the only observation that I have on the matter of experience is the observation of my system, and I cannot say anything about other systems.

The physics of our Universe is deterministic and devoid of free will, and as a result, within the system, it is meaningless: the motion of Existence scattering in space–depth with a cold regularity, just like sand flowing through an hourglass.

An inner consciousness in the system of the hourglass will never understand its significance as a timekeeping system. It can characterize the physical regularity of the flow of the fundamental particle of the system—a sand grain—but not beyond that. Only a consciousness in an external system that contains the system of the hourglass can understand the significance of the system as a system for measuring time.

We see in observations that the significance of every system in the universe is derived from the external system at a lower resolution that contains it. The significance of the Marketing Department is derived from the significance of the company, in which it operates. The significance of the respiratory system is derived from the significance of the human system that contains it. The significance of a molecule is derived from the significance of matter.

A clarification: when I say that System A contains System B, I do not mean in a physical sense, i.e., that it envelops it or surrounds it in space. I mean that it contains it in an algorithmic sense. The system algorithm of System A contains and operates the system algorithm of System B.

Logic (and, unfortunately, in light of the constraints of our internal system, we do not have anything else by means of which to analyze the observations) dictates that the understanding that the significance of any system in the universe is derived from the external system that contains it, must also apply to our Universe system as a whole. If we look at the lowest possible resolution, we will see our entire Universe as one system with one system algorithm—it is said to draw its significance from the external system that contains it.

Be prepared for a surprising twist...

The observation on experience as a meaningful awareness (feeling), has left me no choice but to state the Sixth Axiom:

> *The Sixth Axiom - meaning*
>
> *"There is an external system that contains the internal system of our Universe, and the meaning of our internal system is derived from it."*

I decided that the Sixth Axiom would not be included in the five basic axioms that define our Universe. It remains outside of them, like a step-daughter—as five-plus-one-more. As distinguished from the five basic axioms, the Sixth Axiom does not add any information or insight for us about the physical regularity of our Universe. The only thing you can say about it is – that it exists.

In the context of the Sixth Axiom, I find it appropriate to refine two points:

As consciousness constrained by the internal system, we do not have the possibility of saying anything about the external system, apart from the fact that it exists.

  1. Do not become confused with our intuitive feeling that

the external system physically surrounds or envelops our system. That is to say, that you should not go to the outer limit of the Universe and be disappointed to see that there is nothing there... Just like a very wise Sims person will go in his universe and investigate its regularity—he will not be able to understand that the external system that transcends his universe is just a metal box that contains silicon inside various chips, where information about his hand is stored in Chip A, and the information about his head is stored in Chip B, where it just happens that information about his dog is also stored...

2. We cannot say anything about the external system that contains us, including the way it contains us.

Chapter 27
Is God is a Binary Force?

Tell me who God is, and I'll tell you if He exists.

Another matter that must be treated, when trying to define the physical reality in which we live, is God. What is the essence of this force, and is there such a force that influences the universe system?

Without offending the beliefs of human systems, the prevailing opinion in scientific thinking is that there is no evidence in scientific observations for the existence of an intervening god. There is no evidence to the effect that there exists a force external to the Universe that listens to prayers, except for testimony that come from subjective feelings of human systems. I say this as a very devout person. There is no contradiction between my subjective faith that God, who listens to me, exists, faith that helps me cope with the anxieties of life, and the fact that from a physical standpoint there is no such deity. I am perfectly satisfied with His existence in my imagination. An imagination that there is a god, who assists us, truly assists us, even if this is just by way of facilitating dealing with anxiety. Religious belief has developed as a psychological response to the anxieties felt by human systems as a result of the reality according to which their destiny does not depend on what they do.

In any event, if an intervening deity exists, there is no meaning to the science of physics, in terms of investigating the internal rules of the system. An external entity that changes the rules of the game according to the prayer of one human system or another will cause random changes to the system and loss of regularity. This would be a contradiction to the Third Axiom that states that there is continual regularity.

From the standpoint of General Binary Physics, there is no room for "miracles". The essence of a "miracle" is in creating a particular change in the arrangement of the Existence in a way that does not result from the particular Existence Algorithm in that particular universe or the particular initial state of that universe (it is possible to treat every turn in the queue as an independent initial state). The reason that there is no room for miracles is that all initial states and all Existence Algorithms, without exception, exist in any event in the 7th and 8th dimensions. Therefore, any "miracle" is essentially a description of a state that already exists, from a physical standpoint.

If we refer to God as an internal system that is confined by the Existence Algorithm, Who has the power to create change on a scale greater by several orders of magnitude than other systems—He certainly can exist.

For example, the system of a farmer in Africa has the power to bring about change in his small piece of land, and perhaps among his fellow villagers. The system of the President of the United States has the power to bring about a much greater change. He is able to save a country from starvation, or destroy an entire nation with the push of a button. Of course, it is possible to imagine, in the framework of the axioms, systems that are many times more powerful than the President of the United States in terms of their ability to bring about change. Not only imaginable, but observations show us that throughout the progression in the queue, more powerful systems arise, possessing greater strength to create change.

If we refer to God as a power that is responsible for creating the Universe and programming a framework of axioms, I will say that

He exists. Existence itself is an unequivocal observation that is not in doubt.

Since there is existence, it is possible to make one out of two assumptions about it. First, that it always existed. Second, that there is some force that under some circumstances created existence. Why did He create? Who is He? Or perhaps what is He? Unfortunately, as internal creatures that are confined by the system, we do not have an ability to discover this. We are only able to explore and to discover the internal regularity within the system. Of course, this would be a regularity that was determined by the force that created the system.

In any case, I prefer the second assumption. Although it is impossible to conclude from within the system about things that are happening externally—because our internal system is based on different layers of systems—the assumption that the system that surrounds our system is only a layer within many other systems that create one another, also seems to me more reasonable; thus, there is a starting point in which our specific system was formed. Another thing that hints that there is a starting point at which our specific system was formed is its internal dynamics, a dynamics of development, of which the Big Bang seems to be its starting point. If our system were static, it would be easier to assume that it always existed there, without any starting point. In any event, this is really just my feeling, and there is no possibility of an internal system within the system, as smart as it may be, to infer from within what is going on outside of it.

And as a side remark: There is no need to personify the power that created the system, because any statement that is beyond the confines of the axioms is meaningless.

Despite our limitation in observing beyond the system, I do want to say more about the force that created it.

Since there is a reflection of human systems and their character in any system that they create, I find it reasonable to assume that in our system, there is also a reflection of the force of the Creator and His character. The word "character" is a word that I use having no other choice. I stated before that there is no need to personify the force that created the system. But, since my imaginative ability is limited by the system, as is anything else, I can describe the force of the Creator only, within the limitations of the concepts which are familiar to me.

I believe that even though our system is based solely on algorithms that change cells from a Placeholder to the Existence (And vice versa), qualities such as happiness and love within the system are reflections of physical entities that exist beyond the system. If, for example, the "Sim" game will be further refined and the algorithms will be further developed, and the "Sim" characters will start to become aware of the Universe around them, they will discover an unmistakably Binary Physics, and they will discover that even love and happiness that they feel are based only on clear binary algorithms. But the Sims can certainly imagine that these algorithmic feelings are derived from reflections of similar feelings of physical entities that exist beyond their system… Perhaps it is not just casually written that Man was created in His image…

Therefore, it is logical to assume that the things that are significant in our system are also significant in the system that contains our system. According to the Sixth Axiom, anything meaningful in our system necessarily derives its significance from the system that contains it. We cannot know the weight of the significance that things have in the system that contains us, in relation to the range of things in our system. As internal observers that are confined by our system, we do not have the capability of stating anything about the system that surrounds us, besides the general statement that by the very fact that it surrounds us, the meaning of the things in our system is derived from it.

In conclusion, if you ask me: Who is God? I would answer you that God is a system.

GOD IS A SYSTEM

Each physical system contains all systems below, and gives them new meaning.

The electron is part of the system of the atom.
The atom is a part of the system of the molecule.
The molecule is a part of the system of the cell (for example).
The cell is part of the system of the human system.
The human system is part of the solar system.
The solar system is part of the galaxy system.
The galaxy is part of the system of the Universe system.
And the Universe system is a part of the system of god.

Therefore, it is possible to say that God is found in all things. Every subsystem necessarily derives its meaning from the system above it. That is to say, the uppermost system that we are capable of discerning—the one from which the significance of all observable subsystems is derived—is God.

The Sixth Axiom states that this line of reasoning continues, and even the system of the Universe is surrounded by an external system, from which its significance is derived.

However, because of our limitations and within the settings of the internal system, we—as human systems—are not able to see the external system that surrounds us, or even imagine it.

God is the external system that surrounds our internal system.

And therefore, as for your question – yes, She[2] exists.

---

2. In Hebrew the word "system" is a "female" word, so if God is a system, it is right to treat him as a "she".

Chapter 28
Calculating the Resolution

Because I am a physicist and not a mathematician, this chapter is an anomaly in this book.

But I could not resist including the experience, albeit my preliminary one, of calculating R(h), the resolution of human systems.

It made me very curious to know how many turns there are in one second, and how many cells are contained in each meter. What is the basis of the Binary Field that contains this entire Universe that surrounds me?

I have made the following calculation, which is presented before you.

It should be noted that whether the calculation is right or not, it does not in the slightest affect the insights, the foundations, and the principles of Binary Physics that have been mentioned above. If the calculation is not correct, I am sure that a cleverer person will come who will calculate from scratch in the proper manner.

f = mcl * R (refers to the frequency in depth)

$E = mc^2$
E = f * h (frequency*Planck constant)

As a result,
$f = mc^2 / h$
$f = mcl^* c^2/h$ (refers to f in space)

(We convert m to mcl, because we are speaking of the mass of the cell in depth, not the mass of space)

Equate the two above formulas for f in boldface (assuming that the frequency in depth is symmetrical to the frequency in space) and we receive:

$mcl^* c^2/h = mcl * R$
When we simplify the expression, the result is:
$c^2/h = R$

Hence the resolution of human systems' consciousness, the resolution that is called R(h), is approximately equal to c^2/h; that is, 8.9×10^{16} m/s (the speed of light is 2.99×10^8 m/s; so $c^2 = 8.9 \times 10^{16}$ m/s) divided by Planck constant $h = 6.626 \times 10^{-34}$ J/s. It follows that the resolution of the turn with respect to the second is 1.343×10^{50}.

That is, there are 1.343×10^{50} turns in one second.

I say "second" as the reference unit for R(h), because both c and h refer to the second.

It is also possible to say that the resolution of the observer is a value that is determined to be directly proportional to the value of the maximum response speed, which is the speed of light. That is, if the observer is able to see in one of his mental units (for example, the second) a movement of 1000 cells, then the resolution at which he is observing is 1000.

From the aspect of the resolution, space is equivalent to depth (time) because the response time of both is equal.

When two cells are considered to be one due to the resolution of distance, then the time to go through this "one" cell will also be doubled.

The resolution is essentially the concept of the resolution of observing change: At what resolution do we see change? The resolution of this change is symmetrical in terms of time and space.

When, for example, I say "Resolution 2", I mean that I am referring to any two cells as if they are one cell.

Therefore, space will also be reduced two-fold, and time will be reduced two-fold (because the response time to traverse 2 cells is double that of traversing one cell).

In the fixed universe, $R(0)$, the distance in units of cell is symmetrical to time in units of turn.

$R(0)$ Maximum speed – one cell per turn.
$R(h)$ maximum speed – 3×10^8 meters per second.

So if one second is equal to 1.343×10^{50} turns, and light travels at 299,792,458 meters per second, so $1.343 \times 10^{50} / 2.99 \times 10^8 = 499 \times 10^{39}$.

Thus, in one meter there are 499×10^{39} cells.

An important clarification: Because both the meter and the second are relative concepts, the conversion refers to them being measured on Earth at rest.

Epilogue

Binary Physics is in its infancy.

Generations of physicists will further refine it beyond recognition.

In my opinion at present, work is principally required in the relationship between the Eighth, Ninth, and Tenth rules. To understand and to resolve the way in which they work together. To represent, by computer modeling, a model of a binary atom. To represent, by computer modeling, that the value of π is due to the structure of the cells in the fundamental resolution. And so all the mathematical constants.

Since, throughout the writing of the book, I consulted only with myself, my computing resources amounted to Excel, and I had only three hours a week available (in a good week), I could imagine only the principles. I am certain that a team of researchers with significant computing resources that will run various models on powerful computers, will be able to refine greatly the definition of the rules and the nature of the relationships among them.

I hope that the foundations that were laid in this book will help them interpret the regularity of the Universe…

Understanding the system and how it works will facilitate development of mathematical tricks and searching for lacunae, in order to do amazing things…

Although ultimately as any system, our system as well, limits us as internal systems that are confined by it, still it is a system that allows an unimaginable wealth of possibilities and we, human systems, are very far from exhausting them…

I would prefer that the research will show that there is free will and that the system is not deterministic. Unfortunately, this is not so. I cling to the physical understanding that although I cannot say anything about the external system that surrounds the "internal" system that defines me - that system necessarily exists. I believe that the meaning of our internal system in general, and the meaning of good in particular, is derived from the external system that surrounds it.

And finally, if despite everything that I have written, it is still difficult for you to believe, I will only say that fortunately the truth is not affected by the number of people who believe in it…

Main Sources of Influence and Bibliography

Everything that my system saw and heard, from the turn in which I was born until completing writing this book, especially:

- The Hebrew Wikipedia website
- Studying physics at 5 matriculation units, according to the curriculum of the Israeli Ministry of Education, 1993
- **The Evolution of Physics**, Albert Einstein, Leopold Infeld, Poalim Publishers, 4th Edition, 1978
- Wolfram Alpha website
- **Chaos**, James Glick, published by Ma'ariv - Hed Artzi, 1991
- **A Brief History of Time**, by Stephen W. Hawking published by Ma'ariv, 12th edition, 1990
- **Guns, Germs and Steel**, by Jared Diamond, Am Oved Publishers, 10th edition, 2010
- **The Selfish Gene**, Richard Dawkins, Dvir, 1991
- The film trilogy **"Back to the Future"**, 1985, 1989, 1990
- The film **"The Matrix"**, 1999
- The film **"Cube"**, 1997
- The film **"Cube 2: Hypercube"**, 2002
- **A Brief History of Humankind**, Yuval Noah Harari, Dvir, 2011
- **The History of Tomorrow**, Yuval Noah Harari, Dvir, 2015

- **A New Kind of Science**, Stephen Wolfram, Wolfram Media Publisher, 2002
- **Dante's Equation**, Jane Jensen, Opus Publishers, 2006
- **Blindness**, José Saramago, New Library Editions, 50th edition, 2015
- **Psychological Time**, Dan Zakkai, Ministry of Defense, 1998
- **The Really Important Things**, Haim Shapira, Kinneret-Zimora Bitan, 2009
- **Galileo**—Periodical of Science and Thought
- The monthly periodical National Geographic

www.ingramcontent.com/pod-product-compliance
Lightning Source LLC
Chambersburg PA
CBHW071409180526
45170CB00001B/35